How the West Was
WARMED

How the West Was
WARMED

Responding to Climate Change in the Rockies

Beth Conover, Editor

FULCRUM
GOLDEN, COLORADO

"Climate Revelations" by Auden Schendler appeared, in an earlier version, in *Orion Magazine*

"The Next West" by Stephen Trimble appeared, in an earlier version, in *Portland* magazine

"The Universe on Blacktop (My Day of Saving 66 Million BTUs)" by Laura Pritchett appeared, in a shorter version, in *High Country News*

"Dead Trees" by Jim Robbins appeared, in a shorter version, as "Bark Beetles Kill Millions of Acres of Trees in West," in *The New York Times*

"What's Killing the Aspen? The Signature Tree of the Rockies Is in Trouble" by Michelle Nijhuis appeared, in an earlier version, in *Smithsonian* magazine

"The Big Bonfire—What Colorado Can Learn from the Samso Experiment" by James R. Udall appeared, in a shorter version, as "The Little Island That Could" in *High Country News*

"Homegrown Security" by Chip Ward is a revised version of the article "We Need a Department of Homegrown Security" that appeared in *The Nation*

Library of Congress Cataloging-in-Publication Data

How the West was warmed : responding to climate change in the Rockies /
Beth Conover, editor.
    p. cm. --
 ISBN 978-1-936218-02-8 (pbk.)
 1. Climatic changes--Rocky Mountains Region. 2. Rockt Mountains
Region--Climate. 3. Climatic changes--West (U.S.) 4. West
(U.S.)--Climate. I. Conover, Beth.
  QC903.2.R54H69 2009
  551.6978--dc22

2009036168

Printed in the United States of America on recycled paper by Malloy, Inc.
0 9 8 7 6 5 4 3 2 1

Design by Jack Lenzo
Cover pine beetle image courtesy of cpt.spock/Jay Miller via Flickr® Creative Commons

Fulcrum Publishing
4690 Table Mountain Drive, Suite 100
Golden, Colorado 80403
800-992-2908 • 303-277-1623
www.fulcrumbooks.com

For Jeremy and Ross, of course.

CONTENTS

PART THREE: **COMMERCE AND INDUSTRY**

PART FOUR: **POLICY AND POLITICS**

FOREWORD

The Chinese curse "May you live in interesting times" was perhaps placed on future residents of Colorado by the migrant workers who came to lay railroad track across our plains and over our mountains in the nineteenth century.

These are, indeed, interesting times. When has life in this part of the country ever been dull? When I came to Denver as an exploration geologist in 1981, people said the state bird was the construction crane. The economic boom was strong, and the bust that followed equally dramatic. I remember the day I was laid off from my job and the widespread sense that life would be harder for a while. But a number of us, in a vacant warehouse district and with too much time on our hands, started businesses that led to the broader revitalization of our community. Our first restaurant, in a restored historic warehouse, helped launch the restoration of a place that within a few years was alive with new shops and restaurants and galleries. It became an early sustainable development.

What the Chinese curse doesn't say is that interesting times, more often than not, inspire resourcefulness and creativity that can lead to remarkable opportunities and productive outcomes.

The challenges we face in 2009 are different from those that presented themselves in the mid-1980s. The opportunities are different as well, but they exist. This book is relevant to the challenges and opportunities we face today. This time, the economic downturn in the Rockies follows on the heels of a national recession with international consequences. This time, we understand much more clearly the opportunity—if not the absolute necessity—to plan for environmental costs and benefits in our economic growth strategies. We understand, too, the leadership role that cities and states play in identifying and carrying out solutions and generating entrepreneurial opportunities.

And this time, just as last time, it is the response of the restaurant on the corner that will signal the recovery of the broader economy. The business that reduces its energy bills by taking advantage of emerging state and utility programs that support energy efficiency and renewable energy may find itself at a relative advantage to its neighbors the next time energy prices spike. Investing in these innovations will support related companies that are training workers in "green-collar" jobs such as conducting business energy audits and installing solar panels. In the end, this business likely will realize a well-deserved green marketing advantage as well.

Within city government, greening our processes has also improved the city's bottom line. Greenprint Denver was one of the first programs of its kind in the country. Setting goals and tracking outcomes with specific metrics have helped us gauge specific cost savings at city facilities related to environmental and energy-efficiency improvements, including reduced paper purchases, the citywide installation of LED traffic signals, more energy-efficient lighting at the Colorado Convention Center, and reduced waste at facilities such as Red Rocks Amphitheatre, the Denver Zoo, and the Denver Botanic Gardens.

In 2007, Greenprint Denver staff engaged a diverse and bipartisan group of community, civic, business, and environmental leaders, along with scientists from the University of Colorado at Denver, to undertake a greenhouse gas (GHG) inventory and climate action plan. The result was a goal to reduce the city's GHG emissions by 10 percent per capita by 2012, with an absolute reduction of 25 percent by 2020, consistent with President Barack Obama's goal to reduce emissions nationwide by 80 percent by 2050. Just a few examples of measures now under way as part of this plan are:

- A neighborhood weatherization program launched in 2008 to bring energy and water conservation services to the doors of our citizens. This program is a quadruple win for our community: it saves residents money on their utility bills, better insulates their homes for all Colorado seasons, helps our environment by reducing Denver's carbon emissions, and creates new green jobs by increasing demand for needed home-weatherization services.
- Installation of four substantial solar arrays on municipal property, including a two-megawatt solar array at Denver International Airport and the two largest rooftop solar arrays in Colorado, one at the Colorado Convention Center (300 kilowatts) and one at the Denver Museum of Nature and Science (100 kilowatts).
- An increase in the citywide recycling rate to 69 percent and an organic food-waste compost-collection program for city residents that allows us to divert food waste from the landfill and reduce carbon emissions generated from methane.
- Taking our award-winning municipal green-fleet efforts a step further by

working with the US Department of Treasury and entering into an Eco-Partnership with the City of Chongqing, China, Ford Motor Company, and Changan Auto Group to plan for and procure plug-in hybrid and electric vehicles for both municipal fleets.

- Demonstrating at the 2008 Democratic National Convention in Denver, the greenest national political convention ever, that it is possible to be both pro-environment and pro-business, and that communities can be innovative while also being good environmental stewards. In particular, our Freewheelin Denver bike-sharing program engaged riders from all fifty states and twenty-nine countries in 5,552 bike rides totaling 26,416 miles. This allowed us to avoid 9.2 metric tons of carbon emissions and evolved into a legacy bike-sharing program for the city.

As detailed in the following pages, the Rocky Mountain region is a showcase for both the most immediate and most dramatic impacts of global warming, from the mountain pine beetle epidemic to shrinking glaciers. It is also home to many of the most creative and forward-thinking public and private responses that will help to shape national solutions to this problem. Aspen's ZGreen program, Denver's bike-sharing initiative, New Belgium Brewing Company's corporate culture, and Xcel Energy's evolving mix of energy sources are but a few examples.

Like many before me, I was drawn to Denver because of the opportunities it represented to those who had big ideas and were willing to work hard to realize them. Responding to global warming offers an entirely different set of opportunities, so long as we apply entrepreneurial spirit and perseverance to solving the challenges we face as cities, as a region, and as a country.

In this compelling volume, editor Beth Conover, my good friend and the architect of Greenprint Denver, brings us a wide variety of illuminating perspectives on the challenges and opportunities presented by global warming. From the relevant observations and opinions of local celebrity chef Sean Kelly to those of energy guru James R. Udall, you will see there is no shortage of creativity and ambition in the West to help us respond to one of the great challenges of our lifetime.

There is no single menu theme here, but plenty of food for thought. Sample it. Chew on it. Share it with friends. Enjoy.

—Mayor John Hickenlooper
Denver, Colorado

INTRODUCTION

What does climate change mean for us? What does this world-changing force have to do with our daily lives? How will it change the way we eat, shop, travel, and work? How will it change the landscapes we drive by or the people we interact with every day in some recognizable or understandable way? And what does it have to do specifically with the place where we live? In short, how do we make sense of global warming?

The idea for this book came out of dozens of discussions I had on a daily basis with friends, clients, and acquaintances who felt they understood the basic facts about global warming and carbon emissions and had begun to consider their own carbon footprints at the individual, company, city, or state level, but still had a hard time putting all the pieces together or understanding how the big picture applied to them. With time passing quickly and changes in the physical and political landscapes moving apace, I wanted to create a current snapshot of climate impacts and responses in our region. This book includes a wide range of issues and perspectives—personal and political, technical, cultural, scientific, and geographic—that are arising as individuals and institutions in our region wrestle with climate change and try to respond appropriately.

Global climate change is overwhelming. No other issue in our time has required such a fundamental re-accounting of the energy we use to maintain almost everything we as Americans take for granted as our birthright—our large comfortable homes, safe and easy transportation, industrial infrastructure, and access to a wide variety of reasonably priced, globally sourced goods and materials. Even though we wish global warming would go away and leave us in the comfortable state we enjoyed before Al Gore and his slide show came along, the vast majority of scientists agree that the cumulative impacts of global warming are so dramatic and far-reaching that we need to grapple with the issue immediately and directly.

Given the scale of this phenomenon, it is easier for us to make sense of what is happening globally by understanding and describing its impacts in our own backyard—in my case the Rocky Mountain West. The trouble is, there's very little information being written in real time on this topic. Newspapers are closing their doors in towns and cities small and large across the country, and among the news outlets remaining, science reporters are low on the totem pole. Publications like the independent nonprofit *High Country News* and small-town papers like *The Missoulian* mix with far-flung reports from *The New York Times* and other freelancers to piece together the news that is available. The rest is mostly anecdote and rumor.

From 2003 to 2007, I served as a policy advisor to Denver mayor John Hickenlooper during a period when interest in climate issues exploded. As the architect of Mayor Hickenlooper's Greenprint Denver program, I spent a lot of time studying global warming arguments. Working with scientists at the University of Colorado at Denver, we learned that the average Denver citizen produces the carbon dioxide equivalent of twenty-five metric tons of greenhouse gases per year (roughly equivalent to a solid block of coal eight feet on each side) in order to heat, cool, and power his or her home and business, to get around in cars and other modes of transportation, and to enjoy the wide range of products we consume. This is somewhat greater than the average in other parts of the country because of our region's heavy reliance on coal-based electric-power emissions. Greenhouse gas emissions are so thoroughly tied to fossil-energy use that changing our habits quickly is impossible. Reducing emissions and becoming truly carbon neutral over time will require a complete shift in the way we live, move, think, and work.

Trying to solve any policy issue in the Rocky Mountain West requires a healthy respect for and understanding of the history of the region and its attitudes. Many people have come here, historically or recently, to enjoy the environment and get away from the policies and constraints of more populated coasts. People, in general, do not like to be told what they can and cannot do. Mistrust of public policy runs deep in many areas, though recent polls show that one exception to this may be environmental policy related to quality-of-life factors (see "Red, Blue, and Green—The Western Political Realignment" by Jill Hanauer et al.). The Rocky Mountain West has become new political territory, and its demographics and future trends are being studied carefully by politicians both locally and nationally.

How, then, will the West understand and address the largest collective-action problem of our time? As these essays indicate, national and local public policy is both driven by and drives a wide range of personal, institutional, and marketplace behaviors in ways both expected and unexpected.

The Rocky Mountain West is home to some of the nation's most well-known and beloved national parks and related icons—Old Faithful of Yellowstone, the Colorado River rapids, the diamond face of Longs Peak, Glacier National Park's snowfields, millions of acres of coniferous forests and public lands, and hundreds of miles of seemingly wild and free-flowing rivers filled with rapids and wildlife. Changes in these iconic landscapes—as forests fade and glaciers shrink; as streamflows decline and impact ranchers, anglers, and kayakers; as urban lawns and casinos replace farmland—are emblematic of the changes we face with global warming, and of the challenges that old thinking will not resolve.

And yet, in part because of the West's entrepreneurial spirit and its love of new beginnings, the New Energy Economy of which our new president speaks so hopefully found its beginnings here. From the Rocky Mountain Institute and The Aspen Institute to the National Renewable Energy Laboratory (and its predecessor, the Solar Energy Research Institute), innovative ideas about energy-efficient building techniques and about harnessing the sun, the wind, and the ground's geothermal energy in place of fossil fuels have been under development in this region for decades. As a result, our region has led the way in recent years as interest in these technologies has reemerged, in part because of concern over carbon emissions. If we are to find a way to grow opportunities out of the hardships presented by global warming, you can bet they will find fertile ground in the Rocky Mountain West.

—Beth Conover

PART ONE
CULTURE AND CONSUMPTION

People have long been drawn to the Rocky Mountain region for its riches. From the gold rush to the green rush, newcomers have been lured by mineral wealth, natural beauty, sun and wind, recreational opportunities, and related employment of all kinds. The present-day environmental stewardship ethic of the region is strong. So it's not surprising that as international awareness of climate change and individual carbon footprints has grown, there has been a tangible local response, as individuals install rooftop solar panels, plow up their yards to grow food, take to the streets on bicycles, and choose green products at an unprecedented rate.

In this section, writers from the region describe their personal experiences with climate-related changes in everyday life. From raising backyard chickens to delivering low-impact, high-quality cuisine, and from evangelical attempts to lower emissions to secular efforts to design zero-carbon vacations, these essays examine what's required to reduce individual greenhouse gas emissions, with sometimes humorous, sometimes transformative, and sometimes poignant conclusions.

Aspen Skiing Company's environmental director, Auden Schendler, looks at the spiritual dimension driving some industry and personal practice changes. Veteran reporter Todd Hartman, formerly of the *Rocky Mountain News*, explores the rise of a movement within Colorado's Bible Belt to address climate change from within evangelical churches. Utah-based writer and photographer Stephen Trimble describes responses to climate change as part of a changing regional zeitgeist.

Popular local author Laura Pritchett turns the extractive heritage of the region on its head as she describes the personal and mineral wealth she gleaned in an afternoon of Dumpster diving. David Akerson, a human-rights lawyer, describes why he drew the line at backyard chicken farming in his quest to reduce his family's carbon footprint. Lisa Jones and Mark Eddy write about different approaches to climate-friendly tourism—in Jones's case, via the ultimate low-impact road trip, and in Eddy's, through a journey to one of the planet's rapidly vanishing landscapes. Denver celebrity chef Sean Kelly describes the daily dilemmas faced by an ecologically conscious food professional. Jackson Perrin and Dev Carey, cofounders of the High Desert Center for Sustainable Studies, near Paonia, Colorado, describe their efforts to teach life skills that will help students live more sustainably. Finally, former *Denver Post* columnist and editor Diane Carman describes her eco-consumer leanings as a twenty-first-century iteration of her father's thrifty ways.

CLIMATE REVELATIONS—
GOD, CLIMATE, AND HOPE
By Auden Schendler

One day, a man named Walter Bennett walked into my Aspen, Colorado, office holding a laptop. He was in his mid- to late fifties, with a graying crew cut, wearing khakis and a button-up shirt. He looked like, and described himself as, a West Texas redneck. His younger (second) wife accompanied him, saying little. As we chatted, Walter mentioned that his daughter had just given birth to a baby boy—a grandson. Walter reminded me of the aging, Cheney-esque board members I'd been hoping would die off so we could actually start doing something on climate change. But that was exactly what he wanted to talk about. He set down his laptop and hooked it up to a projector.

"Do you mind if I show you this presentation I've prepared for my senior management?"

"No problem," I said, thinking, *Get me out of here. This is going to hurt.*

I'm a climate guy. I work for a ski resort, Aspen Skiing Company, where my title is sustainability director. In theory, I work to address all aspects of the resort's environmental impact, from weed control to cage-free eggs, from taking calls about new technologies to handling attacks about what a bunch of hypocrites we are. It's fun. I enjoy it. But, to be brutally honest, I don't care that much about those subjects. Twenty years ago, I took my first course in climate science. The news I read today is essentially the same. And I believe two things: first, to quote ABC newsman Bill Blakemore, "climate isn't the story of our time; it's the only story." Second, it seems obvious that a ski resort should both care deeply about climate change and also be in the vanguard of solving it.

Because my job is high profile, people often ask to meet with me about climate, sustainable business, and the environment. That's what Walter Bennett was doing. Walter works for Stihl (pronounced "steel"), the German chain saw manufacturer. We have a partnership with them. They support free-skiing competitions, and we use Stihl saws on our mountains to cut trails. I didn't expect much from the meeting. After all, we're talking about a chain saw manufacturer here. But after Walter got his projector set up, he clicked a button and proceeded to blow my mind.

He had prepared an hour-long multimedia event on climate change, complete with country music overlays, video clips, and charts and graphs, that rivaled any presentation I'd seen from experts in the field, nonprofit heads, and climate PhDs. It got the science exactly right, the challenges, and some of the solutions. Walter's goal was to convince Stihl that it should begin to take action on climate change, in concert with its efforts to develop cleaner-burning chain saws and other power tools.

When Walter was done, I sat in silence. Finally, I asked, "Walter, if you don't mind my asking, what was it that moved a self-professed West Texas redneck to care about climate change at all, let alone try to change an entire corporation's perspective on the issue? You don't really fit the mold of someone who would do this."

Walter said, "Holding my grandchild—holding that little baby in my hands…" His voice trailed off. I thought he was going to cry.

Walter's experience, I believe, is being lived throughout the country, throughout the world, because climate change is a threat the likes of which our society has never seen. Unlike some earlier predictions of doom from environmentalists (the population bomb, for example), this one has uniform scientific agreement. Climate change is happening, and it will get worse. The best science—represented by Rajendra Pachauri of the Intergovernmental Panel on Climate Change and James Hansen at the National Aeronautics and Space Administration—tells us we have to act in the next few years to cut carbon dioxide emissions 80 percent by midcentury or the planet will be unrecognizable by the end of the century.

And yet, somehow, we don't seem to be able to engage this monster adequately. While Aspen Skiing Company has developed a worldwide reputation as a green company, our energy use keeps increasing, despite herculean efforts to reduce it. Not only are other businesses struggling in the same way, but also most of the nations that signed the Kyoto Protocol are missing their targets. Why? Because our society is entirely based on cheap energy. We can't just retool it overnight. Solving climate change is going to be a bitch.

Given the extreme challenges we face in implementing solutions—whether trying to make mass transit work, fixing the problem of existing buildings, building enough renewable energy to power our operations, or driving federal action on climate policy—it's worth asking the question: what will motivate us to actually pull this off? How will we become, and then remain, inspired for the long slog ahead? Because this battle will take not just political will and corporate action; it will require unyielding commitment and dedication on the part of humanity. We need to literally remake society.

We can intellectualize the need for action all we want, but in my experience, in the end our motivation usually comes down to a cliché: our kids and, for want of a better word, our dignity. Journalist Bill Moyers has said, "What we need to match the science of human health is what the ancient Israelites called 'hocma'—the science of the heart...the capacity to see...to feel...and then to act...as if the future depended on you. Believe me, it does."

Moyers, who is an ordained Baptist minister, taps into something positively religious about the possibilities in a grand movement to protect Earth. Climate change offers us something immensely valuable and difficult to find in the modern world: the opportunity to participate in a movement that, in its vastness of scope, can fulfill the universal human need for a sense of meaning in our lives. A climate solution—a world running efficiently on abundant clean energy—by necessity goes a long way toward solving many, if not most, other problems too: poverty, hunger, disease, water supply, equity, solid waste, and on and on.

Climate change doesn't have to scare us. It can inspire us; it is a singular opportunity to remake society in the image of our greatest dreams.

What are those dreams? The concept of an ideal society has been a core element in human thought for all of recorded history.

In 1516, Thomas More wrote about a kingdom called Utopia off the coast of the recently discovered Americas; in doing so, he brought the concept of an ideal society out of the realm of religious faith and the afterlife and into the world of the living. For centuries, that utopian ideal had been called by different names but had always existed in some other world: the Garden of Eden, Paradise, the Land of Cockaigne. More's idea that such a place might exist here on Earth was radical, but it came from the same yearning for meaning and betterment that has always driven human beings to new heights. One of the great and hopeful concepts of human history, it carried itself into the present: from the settling and then founding of America and all its promise; to the vision behind Kennedy's City on a Hill and Johnson's Great Society; to Martin Luther King, who said that he might not get there with us, but he had seen the Promised Land. The absence of that vision is despair.

Barry Lopez has written, "One of the oldest dreams of mankind is to find a dignity that might include all living things. And one of the greatest of human longings must be to bring such dignity to one's own dreams, for each to find his or her own life exemplary in some way." This longing is a fundamental aspect of human experience. In my work, I see it on a daily basis, in people like Walter Bennett, in the hundreds of college graduates looking for work in the field of sustainability, in people all over the world. Recently, I received the following e-mail from Bob Janes, an Alaskan tour guide I met in 2007:

Greetings from Juneau, Auden,

...My interests are being drawn more and more toward the global warming issue (whose aren't?). I am able to involve myself both personally and in a business capacity now and into the future, but am definitely in the dark on a specific course...

Do you believe one can actually find a way to earn a bit of a living in this emerging (crisis?), and at the same time go home at night and let the kids know that something good is being accomplished? My business sense tells me there are many grand opportunities, but the field seems to be a tempting invitation to intrusive species and interests. What is reality? What will stand the test of time?

When you get a chance, Auden, could you drop me a line with some thoughts and possible information links...

Bob

In a note dashed off after work or between tours in the mayhem of a busy day, Bob was asking some of the most basic, consistent, and profound questions humanity has struggled with. And when I tried to pinpoint exactly what Bob was talking about, I ended up with words that didn't square with the biology background I have or the empirical perspective the field of sustainability and climate has historically followed. The words I found to describe Bob's goals came from the religious community—words like *grace*, *dignity*, *redemption*, and *compassion*. And it occurred to me that the environmental, political, and business worlds, in their discussion of climate change and its solutions, have been missing something fundamental.

There have been scores of books published on climate change and sustainable business over the last two decades. Most come from the secular academic or left-leaning environmental community, or they come from the free market–crazed economists at right-wing think tanks. It's either pure science or pure economics. Few of these books address the broader, seemingly glaring point that no such holistically encompassing opportunity as climate change, nothing with so great a promise to achieve universal human goals on so large a scale, has been offered up since the establishment of large, organized religions between 2,000 and 4,000 years ago. The vision of a sustainable society, with its implications for equity, social justice, happiness, meaning, tolerance, and hope, embodies the aspirations of most religious traditions: a way of living at peace with each other, the world, and our consciences; a graceful existence; a framework for a noble life. Most religions originally evolved to meet a basic human need for community, understanding, and mission. Religion, in its original intent, and the sustainability movement seem to be sourced from the same ancient human wellspring.

Is it any wonder, then, that so many have come at sustainability, and in particular the climate struggle, with an almost religious fervor? And that many prominent leaders of this movement—leaders like Al Gore, Sally Bingham, Bill Moyers, and Richard Cizek—are either ordained or educated in theology? Indeed, many critics of environmentalism and the current climate "crusade" point out the avid, zealous enthusiasm behind the movement, as if to say, "What a bunch of wackos."

But religion has been one of the most important forces shaping society throughout history. If there are some very clear parallels between the goals of most religious traditions and the goals of a sustainable society, how is it possible to talk about huge philosophical issues that cut to the core of human desire—like climate change, which threatens the very nature and existence of life on Earth—without talking about...God?

My inquiry into religion and climate change began through conversations with my friend Mark Thomas, who was at the time studying for a degree in theology at Berkeley. Mark once said, "To think God is some old guy sitting in a chair, you'd have to be insane." As a member of no religious practice and a lifelong atheist who always felt religion was absurd, I found the idea liberating. I was guilty of viewing religion in the most simplistic terms.

When I talk about religion, I'm talking about its core founding principles, not what seems to be the bulk of popular modern religious practice in the United States. As Bill McKibben has pointed out, in America the evangelical agenda prominent in politics—with its unwavering focus on gay clergy, same-sex union, and abortion—has very little to do with the original teachings of any religious faith, let alone Christianity, despite the fact that roughly 85 percent of US citizens call themselves Christian. He notes that three-quarters of Americans think the line "God helps those who help themselves" comes from the Bible. But Ben Franklin said it, and the notion actually runs counter to the founding ideas of most religions, which focus explicitly on tolerance and helping the poor.

At the same time, the American religious community—even the most unmoored element—is on board with climate action. Leaders typically cite a biblical mandate regarding stewardship, describe Earth as "God's creation," and note Jesus's commandment to "love thy neighbor as thyself." I believe this represents the beginnings of a seismic shift back toward core principles in religion, not contemporary distractions—a shift toward the original, more humble aspects of the Judeo-Christian tradition, and away from making tax cuts permanent. In a way, this makes sense. As we move out of an unprecedented age of abundance and back into a world of scarcity, we are going to need these ideas of tolerance and human dignity that help people work together and coexist peacefully. We are going to need these ideas to solve climate change.

The sustainability movement, too, is arguably seeing a shift toward "core principles" in the sense that we're less focused on the microscale and the individual (recycling, paper or plastic, self-righteous sport-utility-vehicle hating) and more focused on the collective (solving climate change as a social, economic, spiritual, and environmental effort).

To get a sense of what might be happening on the leading edge of religion—and how this evolution might relate to the climate struggle—I contacted two young progressive religious thinkers: my friend Mark Thomas, now director of mission integration and spiritual care at Providence Hood River Memorial Hospital, and Rabbi David Ingber from New York's Kehilat Romemu congregation. I asked them about Lopez's "dignity that includes all living things." In the process of listening to their responses, it became clear to me that Thomas and Ingber had a particular definition of "God" that informed their whole worldview. Further, it had nothing to do with my simplistic understanding of the idea of God. Let me explain.

Two distinct concepts of God have existed in parallel since the origin of religion. Theologian Marcus Borg explains them. Supernatural theism, he says, imagines God as a personlike being. Pantheism, however, "imagines God and the God-world relationship differently...Rather than imagining God as a personlike being 'out there,' this concept imagines God as the encompassing Spirit in whom everything that is, is."

Both Thomas and Ingber used this latter definition, what Father Thomas Keating, a leading thinker on the subject of contemplative prayer, calls the "is-ness" of the world, or "is-ness without boundaries." In fact, after conversations with Buddhist leaders, Keating came to a description of God they could all agree on: "ultimate reality." In this context you could also define God as what Lao Tzu called the Tao, or, simply, "the sacred."

Similarly, the Talmud says of God, "He is the place of the world; the world is not His place."

When you talk about God as ultimate reality or the sacred, and if you see religion as a way of relating to the world in a dignified way—a broker for grace—then the religion discussion becomes much less charged. Nobody's trying to get you to believe something ridiculous. Instead, we're simply talking about a philosophy of living.

In response to my question about Barry Lopez's "dignity," Ingber and Thomas both described a faith that has the goal of bringing the natural world into harmony with people, bringing the divine to everyday experience. As Ingber writes, "Religion seeks (at its best) to illuminate our eyes, that is to actualize our capacity to realize, apprehend, see (with the eye of Spirit) that there is nothing but G-d,* everywhere, now and always."

The idea of the divinization of the world—of our lives—is a powerful and unifying concept tying together religion and the climate challenge. It means that it doesn't matter what direction we come from; most people, religious or secular or something in between, can agree on common goals. An atheist might be envisioning an ideal society running on renewable energy, and others might have the same vision but see that as the true meaning of "God's will be done" on Earth. Heaven must look like a sustainable society.

And yet, for someone like me, the question is how do you talk about religious ideas, or use words like *grace* and *redemption* and *compassion* in a business context, which is all about return on investment, net operating income, cash flow, and year-on-year growth?

Aspen Skiing Company is a good case study. In 1994, our mission, though unstated, was to make money by selling lift tickets. That's not very inspiring. Our incoming chief executive officer at the time, Pat O'Donnell, tapping into the idea that people's lives are, ultimately, a search for meaning, suggested that people won't happily come to work each day to make money for the boss man. Instead, we needed a set of guiding principles that would be based in values, not profits, though business success could certainly become one of those values. What resulted was a core mission for the company that sounds radical to the point of frooiness: "We provide opportunity for the renewal of the human spirit." Come to work to do that, and suddenly things change. Your mission as a company begins to evolve. We're more successful than ever, but that's in part because we've begun to

---

* Many practicing Jews do not write the name *God* for religious reasons. See www.wisegeek .com/why-do-jews-write-g-d-instead-of-god.htm for more information.

see ourselves, and our mission, differently. Perhaps our role, in part, is providing safe, gratifying work to members of the community, creating fulfilling jobs about which people can be proud. Perhaps business can be *graceful*. If that transition is happening in one corporation, it can happen in others.

And the business community is indeed slowly moving in this direction. It started, in part, with books like Paul Hawken's *The Ecology of Commerce* and his and Amory and Hunter Lovins's *Natural Capitalism*. Their argument was that capitalism is wonderful, but it has never been practiced. We've always discounted the value of the natural (and human) world and the costs of our impacts on it. Making the costs of air pollution, climate change, and fisheries destruction part of the business equation—and recognizing the true value of the natural resources we use as feed stocks—would in fact be a divine act: it would mean the business community finally seeing not just the bottom line, but the *entire world* as sacred. It would mean seeing the dignity of the world, the harm in damaging it, and the *vision* of a sustainable future.

It is there. It has always been there. Can we see it?

There is a movement within many religions called the contemplative tradition. Contemplation, or contemplative prayer, is a form of meditation, the goal of which is to cultivate an understanding and relationship with the divine—the life force, the ultimate reality of the world. That ultimate reality might be a dignity that includes not just all living things, but all things. Father Keating has called the entire contemplative tradition simply "a long and loving look at what is." He's now eighty-five and living at the Saint Benedict Monastery in Old Snowmass, Colorado, not far from Aspen Skiing Company's slopes.

I decided to meet with Keating, a leader in this field, because the practice of contemplation is in effect the same thing as the practice of trying to solve climate change; both are an effort to pursue the divinization, the making sacred of the world and of ourselves. That's couched in religious terms, but pagans like me might simply call that state of grace "global sustainability." It's the same idea, though markedly less poetic.

You could argue that the world today is utterly missing the clarity Keating's contemplation is meant to provide, and that's why we haven't moved more quickly on climate change. You couldn't get farther from what Keating calls a "radical participation" in the reality of the world than, for example, *Star Magazine* and *Us Weekly*. Those magazines—just like a public obsession with sports or video games—simply take our attention off what matters. If the public at large needs a clearer view of the world, so do businesspeople and politicians, who both base decisions on short time frames—quarterly reports or election cycles that are meaningless without any kind of broader worldview for context.

To someone who asks, "I want to establish a relationship with the divine. Can I come to your monastery?" Keating might reply, "You can have that relationship anywhere, and should." My conversation with Keating reminded me of the many phone calls I get from eager, young, well-educated college graduates who desperately want to get into the "sustainability field." My response is that given the scale of the problems, every job must become a sustainability job. So one approach is to look for ways to turn your own position into one that addresses climate change. If every job doesn't become a climate job, we're not going to solve the problem. Even if you work for the worst of the worst—let's say it's ExxonMobil—we need people inside the beast. We need moles. And there isn't a job in the world that doesn't some-how influence the changing climate.

My forays into religious thinking revealed to me, above all, a desire within humanity to live in a dignified world. This is Walter Bennett's vision while holding his grandchild; it's what Bob Janes aspires to when he warms up his truck each morning in Juneau. Their urges, hopes, and desires are the deeply rooted, very powerful forces that have been part of human experience always.

This is a hopeful concept: maybe humans are hardwired to durably engage, participate in, and relish the challenge of solving climate change, because it offers us a shot at just this dignity. And maybe something even better: maybe we can't help but do it.

Auden Schendler is executive director of sustainability at Aspen Skiing Company. His writing has been published in *Harvard Business Review*, the *Los Angeles Times*, *Rock and Ice*, and *Salon.com*, among other places. His book, *Getting Green Done: Hard Truths from the Front Lines of the Sustainability Revolution*, was published in 2009.

THE NEXT WEST
By Stephen Trimble

Three Wests. Three ways of understanding, three shadows cast into our imaginations by the same buttes.

First, unquestionably real, there is the land itself, an extravagance of mountains and deserts and 200 Native cultures still intertwined with the holy Earth, a place that astonishes at every turn. Indian people remember 12,000 years of stories rooted in this homeland. Homeland, heartland. These are the great divisions of the continent that make up the West. Storms carry oceanic moisture from the Pacific first to California and the Pacific Northwest: Cascade Mountains, Sierra Nevada, Central Valley, Coast Ranges. Beyond this green fringe, the storms peter out. This is the *Desert West*: Great Plains, Rocky Mountains, Intermountain West, Desert Southwest, Colorado Plateau.

Second, there is the mythic West, the *Old West*, a place of endless freedom and the lone rider. Of noble but supposedly vanishing Indians silhouetted against red mesas, of cowboys galloping through Marlboro Country. Of frontier justice and Louis L'Amour.

And there is the *New West*, where instead of the neighborly Thank You, Come Again signboard as you leave Monticello, Utah, you find the twenty-first-century warning: Last Espresso For 438 Miles.

On the surface, The New West has a new look. Starter castles and golf courses displace hay fields as the rural West fills in. Feed stores become restaurants serving mahimahi and ostrich, spice-dusted and nut-encrusted, accompanied by esoteric baby greens drizzled with raspberry vinaigrette. More hands reach for white linen napkins to wipe stray olive oil in between bites of focaccia than for a crumpled chamois to scrub the grease and sweat from a working stiff.

New Westerners keep alive the myths of the Old West. They cherish the romance of solitary riders below big skies, frontier heroes fulfilling a national destiny. They name their daughters and sons Sierra and Cheyenne and Cody. The Sundance catalog appeals to baby-boomer consumers who started with Roy Rogers and Hopalong Cassidy and moved on to Robert Redford. Boots and dusters, conchos and cowboy hats carry the Santa Fe, Scottsdale, and Jackson Hole cachet of style, defining the superficial image of the New West.

Westerners are twice more likely to move within a year than Northeasterners, who have old roots in communities and neighborhoods. The children of the Old West have to move—to town, for college and work. The New Westerners, kin to the frontier boomers who moved from mining camp to mining camp, have no history, no connection. They move on to the next dream, from Taos to Vail, from Sedona to Napa.

New Westerners move to these easy places first, drawn by their glitzy surface beauty as by trophy wives, living on the surface. They move to Livingston and Ketchum and Park City to take advantage of the Yellowstone River trout fishery, the Sawtooth Wilderness, and Wasatch powder skiing. They imagine they live in wilderness but bring with them their laptops and, with free Wi-Fi in every brewpub, stay tied to Chicago and New York. For the first time, Westerners live here with no day-to-day reliance on the land for their livelihoods.

This New West is urban, yet another break with our imagined West: more than 80 percent of Westerners today live in cities. More than half of American Indians live in cities, mostly in the West. American communities recently promoted to census status as metropolitan areas now come from the heart of what we have always thought of as the empty quarter of the West—places like Bend, Oregon; Farmington, New Mexico; Idaho Falls, Idaho; Prescott, Arizona; St. George, Utah.

These are "real" western places, once booming with uranium frenzy, timber fortunes, railroad vigor. Now, they are simply growing, desirable destinations in the American Dream.

New professions, new skills supersede Old Western wisdoms. Men and women no longer need to know where to cut a ditch to bring runoff to alfalfa fields, when it's worth following a vein of ore into a mountain, or how to fell a 200-foot western red cedar. Now, it's more important to have a flair for cooking perfect omelets at the SkyRidge Bed and Breakfast. Or a knack for teaching freshly retired baby boomers to fly-fish the Rogue River. Or a gift for pairing ranchette properties with the dreams of refugees from the San Fernando Valley.

The professionals around the log table at that bed-and-breakfast wear cell phones in their holsters. They ride sport-utility vehicles with bumper stickers that say New York, Paris, Aspen, Moab. The regional economy depends more on the Dow Jones average than on the price for beef calves at the autumn auction in the county seat.

Newcomers start here, on their first giddy encounter with the West. Later, they may come to understand the deeper souls of these places. They may even learn enough to move comfortably to the dry plains of eastern Montana or the moonscape Dakota badlands, to sagebrush valleys in Nevada filled with silence, or to slickrock alcoves of Zen simplicity on the Colorado Plateau.

When they reach these rural corners of the West, they encounter places that skipped the twentieth century, where the twenty-first century overlays the nineteenth century. In the language of geologists, the New West lies unconformably over bedrock—the mythic Old West and the arid land itself. The New blankets the Old—with a gap, a disruption of continuity.

This is the future, where dissonant, unconforming western identities begin to blur and blend, where New Westerners energized by espresso join with ranchers and Indian people to create new stories—twenty-first-century stories. And where everyone learns that they depend, ultimately, on spring snowpack in the local watershed.

Change has become our political mantra. In the West, change is more than a slogan. Drought tempers our dreams of booming growth. We can no longer take the land for granted. As the climate warms, as forests move up mountainsides, alpine ecosystems and species move with them—pushed off the summits, with nowhere to go.

My son once suggested to me—unwittingly—how to reconcile these three Wests, how to cope with change as fundamental as the climate-driven shifts in the biogeography of whole mountain ranges and horizon-spanning deserts.

When he was eight, we rolled off along the bike path below Telluride, Colorado, pausing on our road trip to leave the truck for an interval of kid activity. (Yesterday was skateboard day; today is rollerblade day.) Afternoon thundershowers replaced the radiant light and overwrought blue of the Rocky Mountain sky. Across the San Miguel River, past condos and ski lifts, the mountain wall reared above the valley's thread of meadowlands. The clouds softened the greens, the even light mapping the mosaic of forest trees.

I asked Jake to come to the fence and to look across to the oversized mural of forest colors. I started pointing. There: light green aspen. And, see, over there: dark green fir. Below, along the stream, a colonnade of Colorado blue spruce.

He listened. And then, wholly without guile, he said, "Why are you telling me this?"

A second-grader whose passions were soccer and rock and roll, he had seen great stretches of the West on family trips. But he had yet to begin thinking about where he fits into his home.

His question was fundamental, though. There we stood, enfolded by the New West: Telluride Mountain Village, town houses, gondolas, cappuccinos, and all. And I asked Jake to see past the pleasures of the asphalt path to the wildness of the mountain and forest.

Why know these things? Why know that the crinkly twigs from these spruce make the best kindling in the Southern Rockies? That Pleistocene glaciers carved the flat bottom of this valley? That hundreds of these aspen trees connect underground in clones sprouted from a single seed 10,000 years ago—and that climate change threatens the survival of these historic clones? That our hillside drains into the San Miguel River, which flows into the Dolores, which flows into the Colorado, which dies repeatedly in reservoirs, resurrected below each dam, gradually drained of its wildness, as it crosses the continent in its descent to the Sea of Cortez? That the whole watershed is teetering on the edge of drought, endangered, carrying as much change downriver as it does snowmelt?

Just as we learn the nuances of the bodies of our lovers, slowly, tenderly, we walk over the land and listen to its stories to find our way to the heart of the West. To establish a relationship with this land, to live ethically here, we must think about consequences.

Why know the names of trees? Why know anything about forest ecology? Because in the West, we will always live enmeshed in Place. The West is so big, so diverse, so dry, so fragile, and—still—so wild, that we need this understanding to distinguish between the old dualities and ironies: lie from truth, boom from sustainability, dream from desert, reality from romance.

All three Wests have tantalized us ever since those long seventeenth-century winters in Plymouth and Jamestown. Thoreau walked west toward freedom. Huck Finn lit out for the territories. Owen Wister's Virginian rattled across America on the Union Pacific to reinvent himself as the prototypical gunslinger in Wyoming. Even today, the sign on Interstate 70 westbound into Utah's San Rafael Swell still says No Services Next 110 Miles.

American culture uses these three visions of the West for its own purposes, playing on our understanding and our imagining of each. The overlay of romance and myth never rests easily upon the real and storied land. The land gives its names to car dealerships and trailer parks and upstanding member businesses in the chamber of commerce. Rainier Beer. Mount Hood Realty. Smoky Mountain Pizza. Longs Peak Liquors. Sangre de Cristo Dental Clinic.

It is easy to ridicule such infelicitous pairings of the magisterial landscape and the entrepreneurial economy. It's also easy to make the case that every gesture of recognition offered, no matter how small or utilitarian, enhances our relationship with the land.

The future of the West looks grim on many levels. Cities sprawl across open space, wild country clogs with recreationists, small towns turn to services and tourism and transform into generic strips of franchises. Ranchers give up, sell out, and move to the Sunbelt. Their 3,000 acres of pasture become one more grid of 20-acre ranchettes. Change comes to communities as surely as it comes to the global climate.

And, yet, we have the chance to move beyond the paralysis of old antagonists. New Westerners value the same open space honored by generations of families tied to the land. The two dismiss and mistrust each other. But if they can keep talking, if they can cooperate, New Westerners can absorb the best of the Old. The Old can become the New.

If old-timers can resist embitterment while they give up so much, they can teach the newcomers something about roots. We have a brand-new chance here—the positive spin on change coming to all of us. We can create a mix of rural and urban cultures that recognizes the distinctive landscape, fusing the three shadows cast by three Wests into a single image of the future—a Next West, the People's West, where public lands are no longer remote, where 60 million people live in the once lightly settled interior West, and where public lands interfinger with all varieties of private lands. Here, we will redefine the rights of private ownership in a new geography of conservation, in a post-mythic and wise West at peace with the great land as it changes and copes and evolves. The very Earth challenges us when global patterns in climate redefine our home landscape.

When Wallace Stegner asked for the West to create a society to match its scenery, he didn't imagine that both society and scenery were in flux. Westerners, old hands at re-creating themselves, can take the lead in adapting their society to "scenery" losing its glaciers and expanding its deserts. The Next West may be the next America, the next cultural equilibrium, the next wave in redefining the story of becoming native to this still dry, always dry, sunstruck homeland of ours.

Salt Lake City writer and photographer **Stephen Trimble** has published more than twenty books on western wildlands and Native peoples, including *Bargaining for Eden: The Fight for the Last Open Spaces in America*, *Lasting Light: 125 Years of Grand Canyon Photography*, and *The People: Indians of the American Southwest*. His website is www.stephentrimble.net.

SALVATION FOR THE CLIMATE
By Todd Hartman

The pulsating center of evangelical Christianity in the Rocky Mountain West sits on the high plains just east of Colorado's famed Pikes Peak, an area dotted with sleek, isolated business parks and, just visible in the distance, a sprawling mishmash of speedily erected apartments, tract homes, and mini mansions—"a bunch of ugly boxes," to borrow a phrase from an old Eagles tune about the loss of the western frontier. It's from here—the northern fringes of Colorado Springs, a community sewn together by five major military installations, a veritable industry of conservative Christian organizations, and a fading high-tech sector—that New Life Church draws upward of 10,000 people to its massive worship center on Sundays.

Complete with an arena-sized hexagonal sanctuary, a café that serves cappuccinos and breakfast burritos, a bookstore with a Barnes & Noble–like ambiance, and a dizzying array of Sunday schools where children can bounce on giant inflatable obstacle courses, shoot baskets, and wander through jungle-themed hallways, New Life fits the definition of what has come to be known as a mega-church, or, more derisively, a God Mall. Part theme park, part concert venue, part social center, New Life strikes one as a sort of religious Third Place, an enormous Starbucks, but with Bibles near the register and friendly people reading scripture aloud on the couches. The structure itself outsizes everything around it, including a new multiplex movie theater down the street. Were it ever to be abandoned, New Life's shiny campus (save for the cross on its sanctuary roof) could probably be mistaken for the kind of big, glossy regional corporate headquarters scattered throughout the freshly developed exurban dry lands of northern Colorado Springs.

Inside New Life, lighthearted Christian rock music consumes the first thirty-plus minutes of service. Backed by a band bathed in blue, purple, and red lights, a choir of at least 100 bursts out lyrics in praise of Jesus with so much enthusiasm that some members actually bounce up and down as they sing, engaged in a kind of evangelical aerobics. Many in the audience hold their arms aloft and know the songs by heart. With their eyes closed and smiling faces, buoyed by the spiritual energy of the moment, they need not pay attention to the cinema-sized screens showing lyrics that praise the permanence of God and concede the temporal nature of man: "You will remain after the day is gone and things of the Earth have passed," goes one oft-repeated refrain.

Pastor Brady Boyd leads the service. On a Sunday in early 2009, the energetic, clean-cut forty-two-year-old, dressed in a dark sweater vest and charcoal slacks, is helping the congregation work through new developments tied to a still-simmering scandal involving the church's former lead pastor Ted Haggard, who left in late 2006 after indications he may have used drugs and engaged in a homosexual affair. Pastor Brady, with a mix of stoicism and humor, strains to make the case that the episode—which blew up into a national news event because of New Life's and Haggard's high political and religious profile—has strengthened the church and its congregants. He weaves the matter into a message about temptation, which leads into talk of difficult times and the testing that occurs. God, Pastor Brady says, tests us. And it's in the hard times, the times of testing and trials, when we learn the most about Him.

New Life, with its born-again brethren—with their emphasis on the hereafter, leading a godly life, and, as Brady says in an interview, aiding "broken humanity"—may seem an unlikely place to marshal a new movement to fix Earth's broken climate. Indeed, in the popular mind, evangelicals are hardly considered a wellspring of environmental stewardship, although that viewpoint is not entirely fair when one considers the movement's relentless but little-publicized efforts to bring food, clean water, and other necessities to forgotten peoples. Even so, the media stereotype of the movement is largely on target, portraying it as a powerful wing of the Republican Party, one that in recent elections has very publicly pulled the party far to the right on social issues, with its drumbeat emphasis on outlawing abortions and preventing same-sex marriages.

But this hallowed ground is shifting ever so slowly, like an underground fault grinding away, largely out of sight but with an occasional tremor. A new generation of evangelical leadership, including until his travails New Life's former superstar pastor Haggard, as well as small armies of bright-eyed millennials (those born between 1980 and 1995 and still giving the world a fresh look), is embracing a far broader agenda for their church, leaving aside aging obsessions over personal sin for a focus on poverty, social justice, the environment, and, yes, a warming climate. Their motivation stems from two key concepts: that the Earth is God's creation and caring for it is part of a Christian duty outlined at various points in the Bible, and that climate change will likely affect the world's most vulnerable—"the least of us"—first and most severely, and that alleviating such suffering is a Christian calling.

Pastor Brady, himself a post-boomer not too far removed from generations X and Y and the millennials, is by no means unaware of this trend. And he probably can't afford to be. One thing a visitor quickly notices about New Life is its, well, new life: youth is everywhere. Teens, gleaming couples with babies, shiny cadets from the nearby Air Force Academy, and a thick core of thirty- and fortysomethings fill the church's spacious halls. New Life suffers not the problems striking many older, small mainline neighborhood churches, where funerals are chipping away at membership rolls. So in the long run, it cannot ignore what now-blossoming generations have on their mind. If not climate change specifically, then it is stewardship toward the environment and conserving energy that "resonates with millenials," Brady acknowledged. "They have forty or fifty years left on this Earth. They're wondering what's going to happen to my kids? What's my Earth going to look like? They're thinking long-term. They see the future. They're evaluating what their future looks like."

That said, evangelical concern for the climate remains largely out of the reach of the pulpit in the Rockies. It creates some buzz with younger churchgoers and has been the topic of leadership meetings among the staff at places like New Life, but it's gaining its highest profile on the national stage, where several fledgling evangelical eco-movements are trying to wrest the denomination from its laser focus on gays, stem cells, and the unborn. Those shopworn chestnuts of the movement still matter. But new leadership is suggesting that it's time for a host of other concerns to come in from the cold.

In 2006, dozens of evangelical leaders across the country signed a revolutionary statement entitled "Climate Change: An Evangelical Call to Action." The document, now with some 250 signatures from pastors and other church activists, launched evangelicals on an unprecedented campaign of engagement. The statement declared global warming to be an issue of genuine concern, pinned responsibility for the problem on mankind, and called upon Christians to take actions to address it. A portion of the 1,600-word statement (laden with biblical references) said this: "Christians, noting the fact that most of the climate-change problem is human induced, are reminded that when God made humanity he commissioned us to exercise stewardship over the Earth and its creatures. Climate change is the latest evidence of our failure to exercise proper stewardship, and constitutes a critical opportunity for us to do better." The document sites Genesis 1:26-28 to support that particular passage.

The last decade has seen similar developments bubble up across the evangelical landscape. In 2002, the Evangelical Environmental Network (EEN), an organization founded in the early 1990s, launched a campaign called WWJDrive, or What Would Jesus Drive, to encourage wiser transportation choices. Author and evangelical Matthew Sleeth has sparked his own following. A former emergency-room doctor and Massachusetts hospital executive, Sleeth was awakened by seeing so many patients stricken with chronic diseases such as cancer and asthma. Turning to scripture for guidance, he concluded that an overhaul in how we live and treat the environment was in order. He also walked the walk—he sold his large home and gave away more than half his possessions in search of a simpler, less consumer-driven life. Sleeth gained wide notice in religious circles in 2007 with the publication of *Serve God, Save the Planet*, a book outlining his views tying worship to what's increasingly dubbed "creation care."

Younger evangelicals are making their own mark. Restoring Eden, an organization stirring up life among Christians on college campuses, has made "getting serious on ending human-caused global climate change" one of its four public policy goals, along with protecting the Arctic National Wildlife Refuge, strengthening the Endangered Species Act "to restore creation to fruitfulness," and supporting social justice and human rights for indigenous cultures. The group has sent teams of fresh-faced college evangelicals to lobby Congress, a move that gives them entrée to even its most conservative members. In 2006, Restoring Eden teamed with the EEN to promote the Evangelical Youth Climate Initiative, "a declaration by the next generation of Christ followers telling their elected and faith leaders that climate change is a moral and biblical crisis that demands action."

Remarkable shifts have also occurred within more mainstream Evangelicalism. Perhaps the most notable: the push to protect the climate by the National Association of Evangelicals (NAE) and its former spokesman

Richard Cizik. So startling was Cizik's advocacy—sparked by his own real-ization that distorting the ecology of an entire planet failed to show proper concern for God's creation—that he drew an angry backlash in 2007 from the most politically conservative and visible wing of evangelicals, including James Dobson of the Focus on the Family ministry, himself another noisy cog in Colorado Springs' Christian industry. Dobson and several other con-servative evangelical leaders demanded the NAE call for Cizik's resignation because, among other reasons, it's uncertain "why it [climate change] might be happening and what should be done about it." Cizik survived, in part perhaps due to a hard push back from evangelicals believing otherwise. In an open letter, evangelicals from Restoring Eden expressed their disagree-ment with Dobson, accusing him of "shocking haughtiness" and articulating what reads like years of pent-up frustration with the more politically conser-vative wing of the movement. They wrote, in part:

> Many evangelicals have for more than twenty years pled with James Dob-son, Tony Perkins, Gary Bauer, D. James Kennedy, Jerry Falwell and the leaders of several other conservative evangelical ministries to stop ridicul-ing and criticizing how creation-care advocates believe we ought to live in obedience to God, and to instead add to their own ministries the biblical mandate to love and care for our Lord's creation…Further, for the Dobson letter signatories to fail to recognize that care of creation is a vital aspect of valuing and saving human life is a strong indication that these leaders are ignorant of the meaning and scope of humankind's stewardship role regarding the Lord's creation—and of evangelical Christians' egregious failure to address creation's degradation.

As it happened, Cizik was eventually forced to leave the group. In late 2008, he was pressured to leave after public comments indicating his shift-ing views on same-sex relationships and support for civil unions. It was an outcome suggesting that old-guard evangelical conservatives won their battle—but could, taking the long view, be losing their war.

All this is not to suggest a Great Climate Awakening inside the move-ment. It will be no small feat to harness the energy and interest of so many millions of evangelicals wedded to traditional dogma and politics, whether in the Rockies or beyond (one reality check: media interviews conducted at New Life prior to the last presidential election yielded plenty of the requisite

"Obama is a Muslim" commentary). Don Wallace of Denver, a former evangelical who's dabbled in several denominations and is currently focused on the practices of Celtic Christians, calls himself a pioneer in trying to connect evangelicals in the Rocky Mountain region with climate and environmental concerns. He's also a realist. He noted with disappointment row after row of sport-utility vehicles filling the parking lots of mega-churches in Denver's conservative southern suburbs (a similar tessellation manifests itself at New Life). "This doesn't show much commitment," Wallace said.

Pastor George Morrison of Faith Bible Chapel, a church of some 8,000 members just west of Denver, offers what is likely a common view of more traditional evangelicals. Morrison, sixty-one, admits his views are conservative, and he's more likely to take what he calls the FOX News view of global warming. While he respects what he believes are expanding views in the conservative Christian Church to take care of the Earth ("Let's be careful how we use things," he says, "turn off the lights..."), he does not believe climate change is a major topic of concern among his congregants. "I don't see it as a big issue," Morrison says. "There's a lot of skepticism when people say *global warming*." Morrison isn't ready to bite on the argument of a moral imperative to act because a warming planet threatens to drown impoverished peoples. Famines and disasters have happened throughout time, and in biblical epochs as well. "Jesus said it himself, 'You always have the poor with you.' There are always going to be these needs." The needs are big enough just within his congregation, he said. Single parents, unemployment, home loss. "These are real issues, so how much time can you spend talking about this global-warming problem?"

Even Pastor Boyd, though clearly in touch with a growing ecological consciousness inside his own church, is careful to avoid any sense of entanglement with pagan, pantheistic idol worship or New Age notions that have traditionally been a barrier between evangelical Christians and even mainstream environmental groups dismissed by conservative commentators as tree huggers. Explained Boyd: "Here's the way I stand: as followers of Christ, we are to steward the earthly resources that have been given to us. We have a responsibility...to not be wasteful with energy, to use products that can be recycled. Those things are commonsense stewardship issues. Here's where we stop: the church exists for broken humanity primarily, not for broken Earth. It should be a focus, but not a primary focus. We're not Earth worshippers. We're God worshippers."

Some scholars trace Christianity's evolving embrace of the environment, and more recent concerns about climate change, to a watershed event— a 1967 essay by University of California history professor Lynn White Jr., who laid the blame for degradation of air, water, and landscapes at the feet of Christianity. His argument: Christianity, through an adherence to biblical assertions that God told man to subdue the Earth and enjoy dominion over its creatures, has treated the planet as something to be exploited for man's benefit.

While the yawning gap between science and religion, man and nature, has a history that could fill semesters of academic syllabi (see Copernicus, Galileo, Darwin et al.), there is ample reason to believe that—at least at the pragmatic level needed to solve earthly problems—the spiritual and the scientific are increasingly finding common ground. Even very recent times are providing weighty evidence: a burst of ecological enthusiasm sprang from religious groups beginning in the early 1990s, and through creation of organizations like the National Religious Partnership for the Environment they began to fret over what science was learning about man's impact on nature. In 1996, the movement broke the bounds of Sunday school when the EEN stunned hard-right Republican politicians with a press conference criticizing Congress for attempting to roll back the Endangered Species Act. The horse was out of the barn.

Today, many mainline Protestant churches, Jewish groups, and Catholics are deep into programs supporting environmental and climate protections. Born-again Christians—evangelicals—are still behind them, as they sort out their earthly priorities in a period of significant political and demographic churn. As their archconservative older leadership begins to fade and a Net-savvy generation of worshippers grows up in a more global, diverse, tolerant, and technologically proficient society, evangelicals' attention is undoubtedly shifting to a far broader menu of concerns. For confirmation, look no farther than the voting booth. While the evangelical vote played a crucial role in electing Republican George W. Bush to two terms, the trend slipped in 2008, when Barack Obama carved away a meaningful slice of this reliably conservative voting bloc. Again, the millennials were key. One-third of white evangelicals under thirty favored Obama—a doubling of support of that received by Senator John Kerry in 2004.

In Colorado, Obama improved his standing by ten points among white evangelicals. The trend struck even El Paso County, home to New Life Church and long Colorado's most important source of Republican votes with its entrenched conservative Christian and military bloc. Bush carried the county with 64 and 67 percent of the vote in 2000 and 2004 respectively. That dropped a small but notable degree in 2008, when Republican presidential candidate John McCain won 59 percent of the vote. "If you put younger white evangelicals together with black evangelicals, Hispanic evangelicals, progressive evangelicals, and others who are not evangelical, I think there is perhaps a shift in the religious landscape and the political landscape that may be a long-term shift," David Gushee, a professor of Christian ethics at Mercer University in Atlanta, told the Associated Press after the November election.

These results won't escape the attention of leaders at New Life. In a blog posting a day after the election headlined "What happens when your guy loses?" Boyd encouraged his followers to be strong and showed the kind of political sophistication lost on the movement's oft-quoted zealots. "God loves strong believers, but has zero tolerance for mean ones…The world is watching us right now to see if we will attack. Instead, let's choose to serve humanity, not curse them."

It is fair to wonder if the great testing that Boyd referred to in his winter sermon amid new allegations around Haggard is now at hand, and if Boyd and others aren't perhaps looking at the matter too narrowly. A media firestorm over a rogue pastor is one thing. It's quite another to try to harness the will of 6 billion people to solve the toughest collective problem yet to face mankind: global warming.

At a recent welcoming ceremony for New Life's newest members, its newborns, an associate pastor asked God to give the parents the wisdom to raise them well. It was the same morning that Boyd told his congregation that it is in the times of trials and testing when we learn the most from God. Perhaps for evangelical Christendom, we are watching the long unfolding of a teachable moment.

Todd Hartman conducts media relations for the Colorado Governor's Energy Office in Denver. He was a journalist for twenty-four years, much of that in Colorado covering environmental and energy issues, most recently for the now-shuttered but still beloved *Rocky Mountain News*.

THE UNIVERSE ON BLACKTOP
(MY DAY OF SAVING 66 MILLION BTUs)
By Laura Pritchett

One of the reasons I occasionally find myself in a Dumpster in my Colorado town has to do with a woman in India that I've never met. I've seen a photo, though, and that's enough to convince me of the need to climb around dirty diapers and empty cans of cat food. In the photo, this woman is holding her hands out, as if to ask *why*, and her face is splattered in blood.

I'm not the only one in the stinky mess. My two children and a friend, Tim, a geologist and Dumpster-diving guru, are also in the Dumpster, tossing cans over the side, winging them high up so they can arc down with a tinkling clatter. We love Dumpster diving anyway, but today is particularly perfect: we've found lots of huge bags of pop cans, aluminum doors from a construction site, and several pieces of aluminum tubing.

"Aluminum Day," my son declares it, and it wasn't until recently that I understood what this meant, or how important it was.

Aluminum is a funky material. It's the most abundant metal in the Earth's crust, its symbol is *Al*, its atomic number is 13. But the most important thing to know is this: aluminum is too chemically reactive to occur in nature as the free metal, and so it gets "locked in" with other elements, especially bauxite ore. And most important of all, it takes an incredible amount of energy to break those bonds and free it up.

Which means, basically, that when you hold an aluminum can in your hand, you're holding a lot more than the element. You're holding a bit of a strip mine and accompanying tailings piles, a power plant, and a smelter. You're also holding bits of boats and trucks and trains and the fuel required for the transportation, not to mention the aluminum-can and soda-making factories. And there is the human and cultural cost as well—for example, the Adivasi peoples in Orissa, a state in Eastern India that is rich in mineral deposits but poor in honoring the rights of indigenous peoples. A group of peaceable Adivasi women who were protesting the opening of a new bauxite mine to be operated by a Canadian corporation recently got beaten up, hence the bloody picture, and hence my sorrow at seeing aluminum cans in the trash.

But oh, how I hate the preachy and judgmental types. So I'll stick to the numbers. While digging out beer cans from the Dumpster (which sits exactly two feet from a recycling bin), Tim and I consider the facts. Recycled cans take 95 percent less energy than those made from aluminum obtained from bauxite. About thirty aluminum cans are produced from one pound of aluminum, and each aluminum can requires about 3,000 British thermal units (BTUs) to produce it. So, Tim tells me, every two or three cans we recycle basically saves one pound of coal.

All this sounds complicated, but it's not. Not when you look up at the sky and consider what three cans means in the bigger picture of a clean and pure sky.

––––––––

After digging in the Dumpster to retrieve metal that other people have thrown away, my son declares that it's Metal Run Time. He determines this by looking at the enormous pile of various sorts of metals Tim has piled in his backyard.

In fact, we are overdue—there is so much metal here that we have to load up Tim's rickety old flatbed trailer and my rickety pickup truck, and still the vehicles are overflowing with chunks of metal. Our vehicles look like two crazy robots driving down the street, with arms and legs sticking monstrously out, threatening to attack the normal cars that dare to come near.

On the way to the recycling center, my kids chatter on about metals and the prices they bring, which proves that they are already smarter than me (and also sweeter and goofier, as when they say things like, "Mom, you're the best mom, because if you suddenly die, we know how to live out of Dumpsters!").

The dirty and work-worn guys at the metal place chuckle and wave as we drive in. They know us by now, and they think we're weird—*Who is this gaggle of people who keep showing up with all this crazy, miscellaneous metal?* they're wondering. But they like to chat with my kids and show them various interesting machines (the can crusher) or sights (the smashed vehicle, the huge bundled squares of smashed cans).

Tim and the guys start unloading metal, chatting about the weather, and haggling over the purity of certain items. All this will take a while, I know, and I need to occupy the kids, who, for safety reasons, I don't really want wandering among the machinery. I spot a canister of sidewalk chalk on the floor of Tim's car (something he's no doubt retrieved from a Dumpster as well), and the kids and I sit down on the blacktop to draw. We draw pictures of the Earth and sun and stars and comets and shooting stars—we draw and draw until we've got a universe on the blacktop. It looks healthy and bright and beautiful. While we draw, we listen: the roar of the can-crushing machine, the beeping of trucks backing up, men yelling to each other, Tim haggling over the price of clean copper.

Finally, the workers are done: we have 108 pounds of cans (that's about 3,215 of those babies), 400 pounds of scrap aluminum, 10 pounds of clean and dirty copper, 174 pounds of radiators (aluminum/copper mixed), 116 pounds of insulated wire, 26 pounds of number-one single wire, 25 pounds of soft lead, 23 pounds of stainless steel, 30 pounds of yellow brass, and a bunch of batteries.

Assuming that the aluminum would be produced by a coal-fired power system, we have saved 18,000 pounds (9 tons) of carbon dioxide from being released into the air. That's 56 million BTUs. Add to that the copper—about 175 pounds—which saves another 10 million BTUs (and an additional 3,000 pounds of carbon dioxide from being emitted). And we're just talking about the aluminum and copper here; we're not even talking about how much earth would have been stripped, processed, and laid waste in a tailings pile left to further pollute.

On top of this, my kids are holding their first $100 bill, which they're looking at with reverence and awe. Compared to their lemonade stands, this diving business is the better deal.

While they ponder the money, Tim and I look at the drawings of the Earth and the universe. "One ton of aluminum produced from bauxite consumes the energy equivalent of thirty-six barrels of oil," Tim says sweetly, with a touch of sadness in his voice. "That's 197 million BTUs. One pound of coal produced from an average strip mine in Wyoming can produce about 7,500 BTUs. A pound of coal burned produces about 2.5 pounds of carbon dioxide, the primary greenhouse gas…"

On he goes, and some of what he says blurs in my mind, like the edges of our chalk drawing. But enough stays clear: the woman's bloody face, the image of a strip mine, the harsh fact of climate change and the effects it will have, the knowledge that our Earth has only so much to give, and the belief that metal needs to be recycled.

Before we load up and go home, Tim suggests we go diving next Sunday again. Everyone lets out a loud cheer, and so do a mountainside and the sky. At least I like to think so; that's the *real* treasure here. I glance at our picture of the universe on the blacktop and wink at the world. I tell you, it's enough to keep the heart happy.

Laura Pritchett is the author/editor of five books. Her fiction includes the novel *Sky Bridge*, which won the WILLA Fiction Award, and the short story collection *Hell's Bottom, Colorado*, which won the Milkweed National Fiction Prize and the PEN USA Award. She is also the editor/coeditor of three anthologies: *The Pulse of the River*, *Home Land*, and *Going Green: True Tales from Gleaners, Scavengers, and Dumpster Divers*. Pritchett has published over seventy essays and short stories in numerous magazines. She lives in northern Colorado, near the ranch where she was raised.

URBAN CHICKENS—LOCAVORISM AND ITS DISCONTENTS
By David Akerson

Raising chickens in the city is a hot topic in the so-called local food, or "locavore," movement. Locavorism is a recent adjunct to the environmental movement. Locavores seek to minimize their consumption of consumer products that have been grown on environmentally damaging corporate farms and avoid foods that have been grown in Chile or Mexico and shipped to Denver by greenhouse gas–emitting vehicles.

The notion of raising chickens was the latest piece in my family's locavore quest. We had already plowed under our postage-stamp backyard and planted vegetables. This left our two dogs with no real place to poop, but we figured that that problem would sort itself out. We also had beehives. And whenever possible, we shopped at the weekly farmers' markets.

We had taken some significant steps to live more simply, less commercially, to be more responsible for our own consumption. But chickens! Now, that represented eggs, a nice protein addition to our tomatoes and squash. The decision, therefore, felt like a righteous one.

Effecting that decision, however, was an entirely different matter, given the fact that no one I knew had yet tried it. What I did know was that vestigial city ordinances still permit residents to have two hens within Denver city limits. Serendipitously, one day I stumbled across an urban chicken class via an enticing advertisement: "Learn how easy it is to have the reward of eggs and more!" "Easy" comported with my vision of me in a hammock, hens leisurely pecking around my tomato plants, devouring undesirable pests, offering up eggs and the occasional drumstick or two.

So began my journey on a sunny and frigid Sunday afternoon. I love autumn afternoons in Colorado, and the Denver Botanic Gardens at Chatfield was made for days like this. On the historical farm that had been relocated to the park, animals were bustling about. My eyes almost hurt from the brilliant blue skies refracting light on the yellowing prairie grass. Families in minivans were arriving for the nearby corn maze.

I waited with a dozen other strangers in a relocated one-room school-house where we had assembled to learn about raising chickens in the city. The teacher was late. I cupped a coffee with two hands. We squirmed on chairs never meant to deliver comfort.

I scanned the class. We were all at least three generations from any actual knowledge of chicken rearing. For our ancestors, raising chickens was a real and necessary fact of life and probably not so romantic. The animals provided eggs, meat, and fertilizer, and they spread the manure of the other animals as they scratched for grubs and other insects. Today, most people's only contact with chickens is through polystyrene packaging. The benefit of raising one's own chickens had long been eliminated from modern life. This class was poised to restore to us the lost knowledge of our ancestors.

Through a wavy schoolhouse window, I spied a large chicken coop in a large fenced enclosure. I made a quick mental calculation—the neighbor's house would need to go. Thoughts of demolition and perhaps a second mortgage were interrupted by the teacher clambering loudly up the wooden steps in her boots.

She settled in at the lectern uneasily. She was tall, thin, nervous, almost agitated, and I detected a faint facial tic. I resisted the thought, but it was inescapable—she possessed a slight poultryesque quality. Perhaps one could spend too much time with chickens. The teacher explained that she had been a chicken farmer in the foothills of Boulder, Colorado, raising 200 or so chickens of the Araucana breed. She described the birds as if she were describing her children. Some were mischievous and full of verve, others were serious, dutiful soldiers that laid an egg every single day. Her favorite chicken, Chloe, had apparently followed her around like a devoted assistant. The teacher's eyes misted when she told us about Chloe—Chloe was no longer with us. I imagined a stir-fry. She regaled us with stories of poultry wackiness, laughing as she reminisced down chicken lane. In the midst of one laugh, did I hear a cluck? I shook it off.

The teacher proudly circulated pictures of her birds. Araucanas were handsome, although occasionally one sported a tuft of feathers protruding in ridiculous proportions in the shape of muttonchop sideburns. A few others were saddled with absurd Elizabethan collars. Perhaps being an Araucana was no different than most animal species—you do what you have to to attract the opposite sex.

Araucanas produce about five pastel-colored eggs per week. The eggs have yolks of deep yellow, renowned for their taste and nutrition by those who know eggs. She sold most of her inventory to gourmet restaurants for about $5 a dozen, and apparently she made a decent living supplying the gourmands of Boulder—until something happened and then suddenly she didn't anymore. The class didn't go there.

Buoyed by the ebullience of our teacher, the class began to clamor for the chicken nitty-gritty. What do they eat? How much space do they need? Can they be potty trained? How do we get started? What about slaughtering?

"Well, to start with, you need to know that they are quite messy, actually," she warned. "And they are *very* destructive." The emphasis on *very* was accompanied by a tic. It turns out that chickens are big eaters.

"Unless you have a lot of space, you have to presume they will devour everything in their allotted space." A vision of towering tomato plants, heavy with fruit, popped into my mind.

"If you have chickens around anything that is green, any vegetation, they will destroy it," she went on. I looked around the class for confirmation of what I considered to be an alarming fact. The other students seemed to be taking this news in stride.

"Also, you need to be aware that chickens poop. A *lot*," she added, looking us in the eyes to make sure we understood the gravity of the point.

Several hands were raised. It became clear that chickens would reduce my garden to a wasteland of denuded clay and chicken crap in short order.

Hindsight is twenty-twenty, but I wondered why the United States had invented a highly toxic defoliant—Agent Orange—during the Vietnam War to clear the jungles. Apparently, we could have just helicoptered in a few thousand chickens that would have laid waste to the place. Instead of a generation of health problems, we could have cleared out a little space in the lunar landscape of chicken excrement and celebrated our victory in style with a barbecue.

"How did you manage with 200 birds?" someone timidly asked. We all nodded in agreement, having wondered the same thing.

"Chickens are great if you have lots of space," our teacher cheerfully elaborated. She had twenty acres of pasture in those foothills.

Before today, I'd read that chickens were actually beneficial to the land. They are walking pest-control machines, fertilizers, and manure spreaders. Back in the day, family farms needed chickens. They were a vital component of farm health and productivity.

"What about my two dogs?" I asked. "Is it possible that they can learn to live with chickens?" I held my breath for the answer to this question.

"Well, there's only one way to find out," she glibly offered. "Sometimes dogs can learn to live with chickens, and other times, not so much." She had no advice other than to suggest unhelpfully that it would be preferable to raise puppies in the presence of chickens rather than introducing chickens to adult dogs.

The talk of dogs apparently triggered a memory. "This reminds me," she moved on to the next topic, "that one of the important issues to consider in raising urban chickens is that they will attract foxes. When a fox gets into your henhouse, it is unlike anything you've ever seen. Foxes are really sadistic." She shared a few anecdotes, one quite suitably utilizing the word *genocide*.

In fact, mass slaughters were in part the reason that I was striving to become more of a locavore in the first place, and part of my interest in chickens.

In 1999, I accepted a position as prosecutor with the United Nations (UN) International Criminal Tribunal for Rwanda (ICTR). The ICTR investigated and prosecuted the Rwandan genocide of 1994. My wife, Katie, and I, along with our two young daughters, moved to Arusha, Tanzania, where the UN court was located. On my second day in the country, I went to the office and immersed myself in work. Meanwhile, Katie and the girls scoured the town for shops to buy food. There were no markets, much less supermarkets, in Arusha. For the most part, what was available was fresh produce from street vendors. Each of the vendors sold one or two items, obviously grown by them or by someone they knew. One vendor might sell small red bananas and papayas. Another, onions and peanuts. Still another would have apples. Acquiring all of the items on your shopping list may require visits to a dozen vendors in different parts of town. While this was initially a frustration, over time we figured out where and who had the food we wanted, and we came to know them personally.

Poultry was a different arrangement. Our chickens arrived each week from the Chicken Man, who delivered his wares to our house on an old bicycle holding the live birds by their feet, five or six to a handlebar. The inverted, scrawny birds were surprisingly complacent. As slaughter operations go, the process seemed efficient and relatively clean. In anticipation of the Chicken Man's arrival, our Tanzanian housekeeper Mama Jessica*

---

* According to Tanzanian custom, when a woman has a child, she goes by the name "Mama" plus her child's name. Our housekeeper's eldest daughter was named Jessica, thus our housekeeper went by the name Mama Jessica. We never learned her given name.

would boil a large pot of water. It would be roiling by the time the Chicken Man arrived. He would carefully get off his bike—still holding his ten or twelve chickens—and present his inventory. Mama Jessica and Katie would confer and decide on two or three birds. Money changed hands.

At that point, the operation shifted into high gear. With an almost imperceptible twist (perhaps because my eyes were closed), the chickens' necks were wrung, and the animals were beheaded and gutted. Mama Jessica plunged the birds into the water and parboiled the carcasses. When the chickens were removed from the water, the skin peeled off like an overripe banana. The entire process took five minutes. There was no waste: the discarded parts fed the dogs, we ate the meat, and we used the carcasses to make stock. After every bit of fat and flavor had been rendered from the bones, the dogs got those as well. (There is no dog food in Tanzania. The very notion is obscene, given how little people eat.)

The chicken process felt surprisingly wholesome. We were eating meat, a luxury in Tanzania, but we participated in its processing and as a result we didn't take it for granted. When we enjoyed chicken stew or curry, we knew what it took to put it on the table. There was no waste, no chemicals, and no packaging.

The chicken experience was but one aspect of an eye-opening array in Africa that germinated our locavorism. As in most developing countries, there were no garbage trucks patrolling the alleys, emptying garbage cans, in Arusha. In fact, there were no garbage cans. Or alleys. Or even street addresses.

What waste our family produced was burned just outside our front door in our own neighborhood landfill. We saw, smelled, and tasted (?) our trash every day simply by leaving the house. Plastic packing that we threw away on Monday would reappear on Tuesday as smoldering toxic clumps releasing acrid fumes. Metal scraps would rust and decay. Glass would shatter and blacken. In town, we had initially delighted in our ability to purchase day-old *New York Times* reclaimed from passengers on incoming planes at the airport and sold by street hawkers. But the papers too would inevitably end up adding to our personal landfill.

Monitoring our landfill each day naturally began to inform our consumption. Every purchase that came in our front door sooner or later would leave again as waste to be added to our pit.

This, of course, is true in America as well, but as Americans our waste is handled for us—simply plop the plastic trash bag into the garbage can and unseen people at unseen times transport it to unseen landfills. Being intimately involved with our garbage changed our view. We had always viewed trash removal as ordinary, part of the natural order of things. After Tanzania, that convenience now seemed a shade irresponsible.

Beyond waste, there were also startling differences in the value the two societies placed on water. We were one of the few families to have running water in Tanzania, but on many days there was no water to run. Even when the plumbing did work, water had to be boiled and filtered. Toting, boiling, filtering, planning, and conserving—managing our water supply became central to our lives. In times of real shortage, we would fill our bathtub with water as a reserve. Through this lens, we became acutely aware of the utter luxury of water in America: it is safe, instantly available, and in seemingly limitless quantities. In Tanzania, we came to equate water with labor. To waste water one day only meant that the next day's labor would be greater—more toting, boiling, and filtering. Like chicken and trash, our consumption of water was inextricably intertwined with the labor necessary to produce it. We became more careful about using water and more appreciative of water in general.

When we returned to Colorado, Tanzania came home with us. We understood that Tanzania wasn't exceptional; it is how most of the world lives. For a while, we delighted in American conveniences. Not having to worry about drinking water. Trash disappearing in the alley. Chicken hygienically packaged into neatly arranged parts and ready to be cooked. Out-of-season and exotic produce available from all corners of the world. A single vendor of food—the nearby supermarket—aisle upon aisle jammed with products with three layers of polypropylene packing designed to last for decades on the shelf. The sheer number of options seemed pornographic.

In Tanzania, we were personally involved in the labor of producing sufficient quantities of food and water for ourselves and then dealing with the waste we produced as a result of it. After returning to the United States, food and water no longer required any significant labor on our part, and we were no longer confronted by a personal landfill. Now we were rethinking old paradigms. Are supermarkets good? Should we have forty varieties of orange juice and three kinds of avocados available—in January? Should it always be so easy to discard items? Should consumption be so disconnected from its consequences?

These were questions with complicated answers, and we didn't have most of them. But we had a general sense of the direction we wanted to head.

We began to seek out less convenience, to be more connected to our food. We sought a way to reclaim some aspects of Tanzanian village life in Denver. We plowed the backyard to grow food. We raised honeybees for sweetener. We bought as much food as we could in bulk at health-food stores with reusable glass jars.

And then one day I stumbled across the urban-chicken-class advertisement. I remembered the inverted birds on the Chicken Man's bike and Mama Jessica boiling water. Raising chickens seemed a natural addition to our Denver village.

The teacher was finishing. Disconcerted, we gathered up our belongings and the class walked over to the historical farm to inspect the chickens in the historical coop. The fenced enclosure confirmed what she had told us. It was a barren patch of dirt marked with violent gouges. Wooden posts were gnawed. Children were throwing handfuls of grass in the pen; the birds fought over it like piranhas. As much as I wanted to include chickens in our new Denver village, it looked like chickens weren't meant to live urban lives. Chickens on a farm with ample space could be beneficial and symbiotic. Urban chickens destroyed vegetation, created excessive amounts of excrement, and attracted predators.

Urban living has a similar effect on humans and on chickens alike. Put humans in a confined, urban space, and we tend to destroy the vegetation, leaving our allocated space barren. We poop a lot and we create mountains of waste. Because we can't grow much food, we buy packaged food that creates more waste. And we attract predators. On the other hand, give humans an ample allotment of land, and we can live in harmony with it and actually benefit it. We tend to grow things rather than destroy them. We live more simply and less wastefully. Perhaps the real question wasn't whether chickens should be in the city, but rather whether *we* should be in the city.

I decided against raising urban chickens.

David Akerson is an international criminal lawyer and lecturer on genocide, war crimes, and crimes against humanity at the Josef Korbel School of International Studies and the Sturm College of Law at the University of Denver.

CLIMATE TOURISM IN KILIMANJARO
By Mark Eddy

The story jolted us. It was a wake-up call we couldn't ignore. It was late 2005, and my wife and I had just read that some scientists were predicting the glaciers on Kilimanjaro could be gone in as little as fifteen or twenty years. Doing some online research, we found many similar stories, including several that said the glaciers had lost 82 percent of their mass since 1912 and the mountain could be ice-free by 2020.

We'd always talked about hiking to the top of Kili—it was on that jumbled list of travel adventures we keep in our heads—but somehow other trips kept pushing it back to next year, and then the next, and the next.

It wasn't as if my wife and I were living in a cave. We knew climate change was happening and that it would have devastating impacts. But we, like a lot of people, thought it was happening relatively slowly. We tend to think of major Earth changes in terms of lifetimes and not years. And at the time, the estimates of how fast the changes were happening were much more conservative than they are today. These changes, we thought at the time, wouldn't have dramatic impacts for quite a while. We had time to see the glaciers.

But this news was transformational for us. If the glaciers Hemingway made famous were really going to be gone in less than two decades, that meant they were getting smaller all the time. We needed to make climbing Kili a priority instead of something we just talked about.

We researched outfitters that fall and winter, and in early 2006 booked a climb for that September. Through the spring and summer, we hiked Colorado's fourteeners to get our legs and lungs in shape. On September 1, we took off for Tanzania.

The seven-day hike took us into another world. The literature says that hiking the 19,340-foot-tall Kili, Africa's tallest mountain and the largest freestanding mountain in the world, is like hiking from the equator to the North Pole—from tropical rain forests, through a landscape dominated by low-growing shrubs and rock and dust, to snow and glaciers.

From the first day walking through the rain forest, up through the volcanic landscape, we were mesmerized. Dominating the scenery nearly every day was the highest peak on the mountain: Uhuru on the volcano Kibo and the glaciers that clung to its sides.

On Kili, you walk slowly, very slowly. In fact, until you do it, you wouldn't think you could walk that slowly for five and a half days (the day and a half downhill is taken at a sometimes breakneck pace). So on the way up, there's a lot of time to think and talk. I wondered about not only the glaciers, but everything that depended on them. What would happen to the rain forest we'd walked through and the white colobus monkeys, giant ferns, and massive trees that lived there? How would they survive without the water from snow- and glacier melt? How would the farmers who depended on the corn they grew on the flanks of the big mountain to feed their families fare if the crops withered when the glaciers were gone? Would the entire area turn into a moonscape covered with volcanic dust and populated with drought-resistant plants like the upper slopes of the mountain?

I was determined to find out what changes were happening on Kilimanjaro. I had looked at historical photos of the mountain and took note of the size and shape of the various glaciers. During the hike, the glaciers were generally part of the scenery from the middle of the second day on. Whenever they were in view, I would ask Victor, our guide, how the glacier we were seeing compared to what he'd seen before. I wanted to know what the difference was today from the time he first started seeing them as a boy, roughly thirty years ago. His answer was always simple, direct, and to the point: "They are much smaller now. Some years they are a little bigger, but overall they are growing smaller all the time." Then he would point up at one of the glaciers and through the air trace the outline of where it used to be. The glaciers today, according to these air drawings, were indeed much smaller.

I filled my photo card with pictures of the ice, and when we got to the top, I marveled at the towering walls of Furtwängler Glacier in the near distance. The glacier was huge and seemed timeless and indomitable, its bluish ice walls reaching toward the sky. It was only when I returned home that I learned a hole had formed in the middle of Furtwängler, a chasm that reached all the way to the rock below. I read that another ice wall had lost about 10 percent of its 160-foot height since 2002.

We've all come back from trips and talked about how they were life changing. How we were never going to forget how it felt to be in that far-off place, free of our daily chores. But before we know it, the joys and pains of the "real world" take over and that magical feeling we were going to hold near is a distant memory.

But this trip produced a profound change in me. It helped me see what was happening all around the planet in a new way. The changes were happening at a startling pace. I started looking for other travel destinations that might be affected by climate change.

The evidence is clear if you believe the science, and I do.

I learned that the second longest coral reef system in the world, the one that stretches from Cozumel at one tip to the Bay Islands of Honduras on the other, is in danger, as is the biggest one of all, the Great Barrier Reef, off Australia, as well as coral formations all over the world. Higher water temperatures lead to bleaching and dying coral, and acidification is a death sentence too. A 2008 report released by the Global Coral Reef Monitoring Network and the International Coral Reef Initiative said pressures from climate change could reach a tipping point by 2018, sending the entire planet's system of reefs into a steep decline. Another ecosystem that seemed so vast and indestructible is in danger of dying.

From year to year, Colorado is at or near the top of divers per capita. We fly in droves from the mountains to the oceans and wander around the depths marveling at the corals and brilliant fish, lobsters, rays, and other wildlife. If the predictions about climate change come true, it will be a vastly different ocean we are swimming through—an ocean dominated not by a vibrant ecosystem but by the bleached skeleton of dead coral reefs. Communities around the world dependent on the ocean for food and tourism will be devastated.

Several years ago, I took a trip to the Oriente, the Ecuadorian Amazon. This is a place where some of the trees are so big you could place a modest-sized house inside them. Where the jungle is so thick the overwhelming sense is that you are surrounded by a wall of green. Yet the Amazon River basin, which is about the size of the forty-eight contiguous US states and covers parts of eight South American countries, has been strangled by major droughts this decade. Some scientists predict that this basin, home to the largest rain forest in the world, could, as a result of human-caused deforestation, sea-level rise, increased temperatures, and drought, dramatically shrink and give way to grassy savannahs.

A forest system so vast and diverse that it is home to tribes of people who still haven't had contact with the outside world, and that, according to the World Wildlife Fund, is home to an estimated 40,000 plant species, 427 species of mammals, 1,294 species of birds, 378 species of reptiles, 427 species of amphibians, 3,000 different types of fishes, and more than a million species of insects, and is so wet that mold will grow on your skin, could turn into a grassland in mere decades.

Right in our own backyard, climate is changing the forests dramatically. Once-green stands of pine are rusty brown, done in by bark beetles that historically were killed by frigid winter temperatures. But now winter temperatures seldom are cold enough, and instead of living one season they last for two, multiplying twice as fast and doing incredible damage to already weakened trees.

Who knows what the landscape will look like in five years, when the trees have gone from green to brown to gray. Sections of forests in Colorado were closed in 2008 when a hiker was injured by a falling tree. The dead pines become widow makers, capable of falling at any time, killing and injuring unsuspecting passersby.

As the trees disappear, runoff will increase, sending sediment into pristine streams and destroying fisheries. Some of the best trout fishing in the world is found in the Rocky Mountains, home to cutthroat, rainbow, and brown trout.

Nearly every day there are new reports that projections of how quickly climate change will drastically alter our planet were conservative to a fault. More people are discovering—just as my wife and I did back in 2005—that some of those places we've taken for granted will be changed forever in a matter of decades.

But what damage are we doing by racing off to see these places before they're gone? The very act of traveling puts more carbon into the atmosphere and hastens the process that is destroying them. As I thought about the ethics of travel, I started wondering about what might be happening in my own backyard.

I cast about for more information, and another article caught my eye. Scientists are asking all of us to become researchers, to start documenting what is happening in our own neighborhoods—when the trees bud out, birds come back, flowers bloom, grasses turn green; what lives and dies and when. Birds are showing up in places where they've never lived before, and animals, plants, and insects are changing their behavior in an attempt to adapt to a new world. It's all happening right where we each of us lives. I recently became a beekeeper, and I am tracking everything they do and all the happenings that affect them.

There's a world of change happening right outside my house. The bees are my barometer. And while I'll continue to travel, buying carbon offsets and trying to reduce my footprint as much as possible, I've embarked on a journey to discover what climate change is doing to one little piece of the planet in Denver, Colorado.

Mark Eddy is a former environment writer for *The Denver Post* and current principal at Mark Eddy Communications, a Denver-based consulting firm specializing in strategic communications. In his spare time, he travels, hikes, and bikes with his wife, Diane, and dog, Charley—and tends his bees. He can be reached at wmmarkeddy@gmail.com.

TAKING VERN HOME—
A CARBON-NEUTRAL ECO-TRAVEL JOURNAL
By Lisa Jones

In the early 1990s, I was the eco-travel editor for a wonderful magazine named *Buzzworm*. It was a plum of a job—every couple of months, I'd do something like fly off to Botswana, where I'd ride an elephant around the Kalahari Desert or to the Yukon, where I'd raft down a glacial river. I did these things in the company of people who valued nature and local communities; ecotourism proponents offered tourists an alternative to spending their money at, say, air-conditioned luxury hotels built on what had formerly been mangrove swamps. They favored tour operators whose nature lodge or canyoneering franchise was owned and staffed by locals, and which ideally made its clients fall in love with (and ultimately want to protect) the natural setting in which it was based.

For years, I had a blast. I got to go to places I would never have otherwise visited. I got to hike over lava floes and chat with bushmen. But over time, I got tired of being constantly in motion. I started thinking that the little town I lived in—Paonia, Colorado—was the prettiest place on Earth. I started thinking that backpacking in the Rockies and gardening in my backyard provided all the excitement I would ever need. And I started thinking about the 800-pound gorilla of international ecotourism: Americans tend to travel by airplane, and airplanes release an awful lot of carbon into the atmosphere. I started thinking, couldn't we all just stay home and behave?

So when *Buzzworm* went out of business, I spent many years traveling not very far from home at all. Which felt just fine; you can have as exhilarating a vacation in western Colorado as you can have in Madagascar. I could bike or hike straight into the gorgeous Rockies from my front door, neatly avoiding the complicated business of carbon offsets (in which the pollution that traveling generates is compensated for by donations we make to clean-power generators). And if I wanted to take a longer trip, I could do what Jack Kerouac did: I could hitchhike. But I didn't know how many forms hitchhiking could take until my boyfriend and I hit the road early in 1997.

Dev and I were sitting on a concrete bench in front of the supermarket in Needles, California, working our way through six avocados, a jar of salsa, and a loaf of bread. We chewed silently and stared at the sun-blasted parking lot. An old man emerged slowly from a battleship-gray, old-model sedan. He unfolded a walker and carefully made his way past a gleaming row of Winnebagos. Eventually, he reached the square of shade that covered our bench. He set his walker aside and sat down. He looked well over eighty. He wheezed. His eyeglasses sat at an awkward angle on his face. He started, softly, to hum. We ate. He hummed. The recreational vehicles came and went. The sun brightened a notch, and I felt the contraction in my eyes as my pupils set themselves on minimum aperture.

"Hi," said Dev.

"Hi," said the man, introducing himself as Vern.

Would he like a sandwich? asked Dev. Nope, Vern said. What he would like is a way out of this town. This state. Goddamn drug addicts everywhere. Thieves. Mexicans. Bad doctors. Opticians. Hippies. Liars.

Vern's eye doctor back home in Gulfport, Mississippi, had sent him to another eye doctor in Needles. To have his cataracts removed. Two months ago. He was legally blind.

But he had a car?

Yep. He met people who needed rides. They drove his car, with him in it, where he needed to go. And now it was time to go home. Back to Gulfport.

But he hadn't had his operation?

Right. The doctor kept putting me back, putting me back.

Why?

Because he's a goddamn liar, and I'm fed up.

Dev and I were full of many things. We were full of cold water and ripe avocados, which, after a week camping in the desert washes of the Chemehuevi Mountains Wilderness, made us feel optimistic. We were full of triumph at having sprung ourselves from our desk jobs to hitchhike south and be hobos for the winter. We were full of a feeling of rightness that we would burn no fuel for the next two months, easily meeting our needs via the excesses of our countrymen. We had started with only $20 in our pockets— no credit cards—figuring with our shared Presbyterian frame on the world that we could get odd jobs when we needed more money. We had already spent one evening bussing tables at a restaurant in Needles.

We thought about Vern's idea. If we drove him 100 miles down the road to Quartzsite, Arizona, he would at least be on Interstate 10, albeit 1,700 miles east of home. There was a truck stop in Quartzsite, and maybe he could find a driver there to take him on to Gulfport. In return for us taking him there, he'd buy us a hotel room for the night. We badly needed a shower. We'd hitchhiked here from Colorado two weeks ago, spending our first night on a snow-streaked mound of dirt behind a gas station in Saint George, Utah, where the nighttime low reached 17°F. We'd spent the next night somewhere south of Las Vegas underneath a mile-long tunnel of barbed wire that had been used for training troops during World War II. For the last eight days, we'd been in Trampas Wash, in the company of edgy wild burros and phainopepla birds, which looked like they were made of black velvet and sang sweet piping melodies suggestive of wet, green, medieval Europe, not the ageless, bony, empty New World. We'd survived on rice and beans and drunk from a tank holding water for mountain sheep that were being reintroduced to the area by the government.

The deal was struck. We would drive Vern to Arizona. My heart flared with hope. Dev was an environmental saint. He had built a cabin out of recycled materials for less than $900. He cut wood and hauled water. I had rented out the house I owned in town and moved in with him earlier that month. He was tall and wise. I wanted to be more like him—despite my supposed credibility as an environmental journalist, I still occasionally drove away from the cabin to the gym twenty-five miles down the road to work out. I wanted to be better, more self-sacrificing, more helpful, less consumptive, more Dev-like. This trip, I figured, would help me be better. The old man was a godsend.

We set out for Quartzsite, me behind the wheel of Vern's 1981 Thunderbird. It was a terrible car. Its steering wheel had only the most casual relationship with the direction of travel. Its red velour ceiling sagged into my hair, rubbing and making crackling static noises until I wanted to howl like a dog.

Vern told us he had worked as a trucker from the 1950s until the 1970s. He loved driving. He had never contemplated giving up his car, even though he was now legally blind. Astonishingly enough, his own brand of inverted hitchhiking worked most of the time, although a few weeks ago a guy had pulled the Thunderbird over in the middle of the desert, said he needed $20 to make a phone call, and once Vern handed it over, walked away. Vern had waited at the side of the road for two days until he was rescued by someone else.

From the backseat, Dev described the landscape for Vern, the mountains and highway turned tangerine in the sunset. The skinny ocotillo cactus looked, he said, like huge spiders. As the light failed, the far-off Chocolate Mountains really looked like chocolate. Then he named the constellations as they appeared: the Big Dipper, Cassiopeia, Scorpio. Vern seemed to enjoy it. I did too. I thought, *That lovely man. That Dev. His quiet gifts.*

When we pulled off the highway, Quartzsite was booming. Every February, it holds a gem and mineral show that brings in hundreds of thousands of people to the heart of town—a fast-food strip right off the highway. We inched through a traffic jam, got into a truck stop called Ted's Bull Pen, installed Vern at the coffee counter, and hovered by the door, watching. He took three minutes to walk to the bathroom. After he finished his cup of coffee, he took a $20 bill from his clip and held it up to his eyes as if it were a washrag he was going to use to clean his face. But he was reading it, ascertaining its value. Then he peered around, blinking. He was surrounded by truckers and gem and mineral enthusiasts, who with each passing moment looked younger, larger, and more capable of violence. The more crowded the place got, the more Vern resembled an orphaned baby bird. Dev and I went to the parking lot and had an urgent discussion. Even though all our stuff was back in our camp, we would leave it and take Vern to Mississippi. Even though we'd be driving a terrible gas guzzler of a car, that car would be driven back to Mississippi whether we drove him or not. Our environmental consciences salved, our mutual love of adventure enlivened, we committed to the trip.

We got back into the car and headed for Mississippi.

Which is when the trouble began. Vern showed us that just because he was a vulnerable and frail old man, he was not a sweet one. I was the only driver, as Dev had left his glasses back at camp. My eyesight, although better than Dev's, required prescription glasses. My regular glasses were at camp; all I had with me were sunglasses. Vern criticized me, catcalled me, all the way from Arizona to the outskirts of Houston, when he quieted down for a while.

Then I looked at the dashboard.

"Dev," I husked as quietly as I could, "the gas light."

Vern may have been blind, but his ears were as sharp as an owl's. The abuse started again. Dev remained quiet.

I made it to a gas station. Shaking with anger and frustration, I got out of the car and ran inside. I went to the bathroom, then, zombielike, walked to the magazine rack and read a good bit of *People* while Dev—who had filled the car with gas and was holding some bills from Vern with which to pay for it—buried himself in a basketball magazine. We eventually got in line to pay for the gas, feeling a momentary sense of solidarity, chatting softly and rolling our eyes, but miles from being able to laugh about the situation.

Then we walked back outside to Vern, who stood there like a recurring nightmare. We'd left him alone for nearly twenty minutes. Terrible in his frailty, he accused us of abandoning him. He stuck out his hand for the change. He accused us of stealing his money. He said he couldn't do everything for us, pay the hotel bills and feed us too.

Dev said nothing.

Electricity skipped across my scalp.

This guy was everything that was wrong with America—mean, ungrateful, misogynistic, blind in every way I could think of, and clinging to, of all things, this horrible car.

"Fuck, Vern," I growled, "I could walk off and leave you right here."

A silence. Then a short, surreal ballet commenced: Vern circled slowly around us with his walker. Around and around he went. Around me, Dev, and the gas pumps, slowly and helplessly, like some crazy model of the moon circling the Earth. Dev and I sat like two twin planets, Dev's head between his hands, his eyes hidden from view. My eyes were wide open, following every step of Vern's progress. My heart boomed with the prayer that he wouldn't have a heart attack. I thought how he'd told me he'd married his wife on their third date. How he hadn't done much to save his money, spending his earnings on family vacations. How his wife and his only child, a daughter, had died forty years before, within just a few days of each other. How, earlier that very evening, he'd chuckled when I started whistling "Moon River."

He wasn't a bad man, I thought. He's just scared.

Dev turned his blanched face toward me and said Vern only had $48 left.

I breathed one, two, three times.

"I'm sorry I yelled, Vern," I said. "I am really sorry. I promise I won't do it again."

He slowed his orbit and stopped in front of me. "I've been hurt so much," he said.

Dear me. Back when I was an eco-travel editor, I would take trips to exotic locales and have conversations with the people I met there. I talked to homeless Native Hawaiians on the beach outside Hilo. I chatted about the difficult parts of history with Navajo horsepackers. Back then, I always felt happy to have contact with the locals, even if they had some difficult things to say. But I'd have done anything to be buffered from this old man at that moment. The trip Dev and I were taking was low-impact from a carbon perspective, but emotionally speaking, it was embedding a large footprint right on my heart.

Oh, well. We were pretending to be homeless and poor, pretending so effectively that there was no way out. We pressed on and got Vern back to Gulfport, back to the Evergreen Apartments, a cluster of one-room green-asphalt-shingled cabins with kitchenettes. There was a vacancy, but true to the spirit of the past few days, it was only open because Vern's only remaining friend in the place had just died.

Still, Vern perked up quite a bit when we reached Gulfport, a complex of glistening beachfront casinos fronting a mishmash of shabby backstreets, apartment complexes like Vern's, and ponds containing half-drowned televisions, decapitated dolls, and incongruous green herons rising out of the muck to flap their ways to distant trees. He had us drive him to his bank, where he had a talk with the vice president and took out a small loan.

Outside, he laid four new $20 bills in Dev's hand. He said, "You got me here," with what we later realized was surprise. He had expected us to rob him the whole time.

"Good luck, Vern," we said. Then we walked onto the beach, scattering gulls into the steel-colored sky. The sand was glorious, immaculate, as long and wide as an airport runway. We walked in silence, feeling Vern's absence sweep over us like a high-pressure system, feeling our synapses take up the positions of our youth, of the time before this trip compressed us into two mincing, shocked adversaries in a casino town far from home.

Vern was part hero, part antihero. He certainly was a relic from another time, from a time when gas cost $0.25 a gallon. An old trucker who couldn't stop truckin'. A man who had lost everything, including his sight and quite possibly his senses, but who refused to give up his car. But mostly he was just a sad old man.

Dev and I hitched out of Gulfport a couple of days later, competing with a host of hitchhikers heading west to New Orleans and Mardi Gras.

Standing on the side of Interstate 10 for hours, we sang "Amazing Grace," riffing and harmonizing in our mediocre voices, becoming completely swept up by the beauty of the song.

Soon enough, a black man in a red pickup pulled over. He was a preacher and also worked shining shoes at the Houston Astrodome, which was where he was heading. We sadly bade him good-bye at a gas station, giving him our last $5 for gas. Not two minutes later, a passing man handed Dev a $10 bill, saying, "Looks like you could use this."

At the next rest area, we were picked up by a jocular heavy-equipment salesman from Illinois. Over the next two days, he drove us all the way back to Needles, buying us a steak dinner and a motel room along the way. He was transfixed by our story, our experiment. He thought it was amazing that we were hoofing it out here without any money or credit cards, me with my former job as an editor, Dev with his PhD in ecology. And as the hours passed and we told him more, we were happy.

This was Dev's ideal vacation. He didn't much go for the kind that involved planes and hotel rooms; he looked like he was being poked with electrodes when we even drove sixty miles from home to get to a really good kayaking run. I was a sucker for a nice hotel room and a worthy kayaking run, and Dev and I broke up not long after we got home. But we stayed friends, and in the ensuing years we'd often talk about our trip south. Although I haven't done anything like it since, the time spent during those two months and thirty-four hitchhiked rides I shared with Dev was the most alive I'd felt in a long, long time.

Lisa Jones's first book, *Broken: A Love Story*, the story of her friendship with quadriplegic Northern Arapaho horse gentler and traditional healer Stanford Addison, was published by Scribner in May 2009. She has written for *High Country News*, *Smithsonian*, *Tin House*, the *New York Times Magazine*, the *Summit County Journal*, the *Burlington Free Press*, and the *Tico Times* of San Jose, Costa Rica. She lives in Colorado with her husband and cat. Her website is www.lisajoneswrites.com.

COOKING FOR THE CLIMATE
By Sean Kelly

During a recent restaurant industry food show, I was enthusiastically informed by a wide-grinning presenter that his chicken products were available in over seventy different specification models, custom designed to fit my every chicken dream and wish. He had chicken fingers and chicken tenders and popcorn this and jerk that. When I told him that I was only interested in old specification model number one—a whole bird, drug free, unprocessed beyond having been killed, plucked, and iced, and delivered to the back door of my restaurant kitchen—he stepped back, adjusted his ball cap, rubbed his face up and down, and slowly started to say that unless we were talking about a whole lot of birds, it might not make any sense for him. He was a processor, a middleman, and couldn't make much revenue on unprocessed birds. I gave him a dismissive chuckle and waved him off. I knew his story before he had even started telling it to me.

Having been a chef for more than thirty years, I've heard countless sales pitches and been faced with thousands of food-purchasing choices. I've also seen the status of chef go from something roughly akin to picking up people's trash for a career to something as elevated as celebrity, entertainer, and—most recently and most importantly—expert. As food has become more complex and restaurants have become more central to human experience, the public looks to chefs for answers not only about how to prepare food, but about which foods to purchase. As a chef, I maintain a responsibility to myself and my community to make—and therefore defend—the food choice decisions inside my business. The public seems more interested than ever before about how these decisions are made.

At a time when there have never been more choices, more products, more information, and more interest in our food as a nation, it has never been more difficult to navigate the aisles of the supermarket or to put together a menu for a restaurant than it is now—because our food has never before been more manipulated than it is today. I have witnessed vast changes related to food over the last quarter century, and the choices aren't necessarily getting any easier. In fact, the more you know, the more difficult it can be to make responsible decisions about high-quality food. Although we talk as a culture about a return to the wholesome ways of feeding ourselves, we seem to drift from that goal as if there were some huge gravitational pull holding invisible powers over us. The powers, of course, are corporate, and they exist to profit from our hectic, fast-paced, dual-income family structures, challenging us to put healthy food in our bodies on a regular basis. And, at times, the government's role in the regulation of our food supply seems to generate even more confusion.

The seventy different types of chicken products are just one example of the direction our food industry has gone in recent years. Whether it's processed chicken or whittled-down "baby" carrots for salad bars, producers want to process the goods they sell. In the boardrooms of the industry, this is referred to as product enhancement, and it's designed to make life easier for the consumer—kind of like doing a favor for pennies on the pound. When the volume of pounds is large enough, a lot of extra pennies add up for the producer. Succumbing to economic pressures on many fronts, we consumers readily accept these favors and outsource much of our food preparation, even when we are "cooking" in our own homes and restaurants. The truth is, though, that the more food is processed, the greater likelihood it has traveled around the country (or world), propelled by fuel consumption, in order to reach us. In addition, foods that travel need added chemicals and preservatives to keep them "fresh." We have allowed far too much of our food to become some sort of scientific hybrid of what it used to be. Much of it is chemically treated to be processed in one hemisphere and consumed in another, but marketed, disingenuously, to appear as though it had all come from around the corner at Grandma's house.

Nowadays, when people ask me what kind of food I serve in my restaurant, they are often asking not only about the style of cuisine, but whether I offer organic or conventional options. This is a question that never fails to unsettle me somehow, because there just doesn't seem to be a right answer for everyone. It's a question that runs along the same lines as where one stands in his political affiliation. The term *organic* has become so popular that it has achieved its own cult status. But corporate organics have evolved over the last decade and diluted the value of the word, leading to a wholesale reassessment of what *organic* is really supposed to stand for, and consequently fracturing a band of once-loyal, like-minded followers.

American food companies, in many cases, have moved "organic" production operations outside of our borders in order to reap the benefits not only of cheaper land and labor, but, more importantly, of lax environmental standards and enforcement, thus creating a situation in which our foods have to travel far greater distances in order to get to us. This, of course, may create a greater environmental threat (in the form of global warming, for example) than the purchase and use of conventionally grown foods that are raised locally. And because organic standards are not universal, as we import more and more of our everyday products from afar, questions of authenticity can arise.

The organic choice is also complicated by economics. Because we have allowed our food industry politics to be controlled by massive conglomerates, we are left with a mainstream marketplace that is primarily devoid of anything resembling quality and wholesomeness at the commodity level. With taxpayer-supplied subsidies, our government regularly assists in the manipulation of a supply-and-demand market, perpetually keeping supply-side prices so low that, regrettably, selecting commodity-grade, conventionally grown products becomes the only viable choice a chef can make in order to keep a business profitable in today's economic climate.

Now, for many chefs, ignorance is bliss, and if their clientele doesn't care, then choosing foods that are conventionally grown and industrially processed may be a fine arrangement. But if you're a chef who doesn't want to be ignorant of the kind of demonic practices that our current American standards allow in the raising and processing of the foods we citizens consume daily, then the situation can feel like being between a rock and a hard place. Restaurants, of course, are public and interactive, surviving on participation and engagement. No chef or restaurant operator can do very much without customers who are understanding and supportive of the concept. In order for that concept to be nurtured by enough customers to keep the restaurant viable, the menu has to be somewhat mainstream. The chef who wants to be devoted to organic foods in his or her restaurant needs to go off the grid, so to speak, in order to find the majority of the food supply necessary, and he will pay more for those foods. Ultimately, that chef has to attract an elite, forward-thinking clientele who will support these actions. That's no easy task.

To move away from mainstream, commodity-level foods to a position of quality over quantity is not without drawbacks. Every chef I know who has been intelligent and passionate enough to strive to serve his or her guests nothing but the most politically correct and wholesome foods is subject to being called a fanatic, because doing so knowingly excludes the majority of potential clients simply due to the higher prices that must be charged. Obviously, except in the most elite of neighborhoods, this diminishes the business's chances for long-term profitability and survival. Knowing the inherent risks involved, setting up a restaurant with a tight, confined, organic-only business plan would seem to border on lunacy. On the other hand, for people to pay absolutely no attention to where food comes from, how it is produced, and what becomes of the land after it is produced is equally bordering on lunacy. It is within this precarious place that the chef—and the conscientious consumer—struggle to make responsible choices.

While there are no easy answers, there are some practices that can help keep us on course.

- Commitment to organics is laudable. In order to take care of our own bodies and our struggling planet, we cannot afford to be ignorant of the facts of rampant pharmaceutical use in our livestock and devastating pesticide practices on our soils. However, when the organic choice involves products shipped from distant regions or products that are overprocessed and overpackaged (such as organic TV dinners or organic frozen pizza from Italy), the meaning of organic is lost.
- When possible, buy local. Go to your neighborhood farmers' market. Meet a farmer. Join a community-supported agriculture program or co-op. By extension, this means eat seasonally. Peaches in January, in the Northern Hemisphere, are coming to us from Chile, where they are picked green, shipped unripe, and put in gassing chambers to ripen—all at a big expense to the environment and to the peach's natural flavor.
- Support a local butcher rather than buying all meat precut and prepackaged from a big-box supermarket. Engage in dialogue regarding the meat, its origins, and its route to you. Don't hesitate to ask questions and discuss your standards. If it costs a little more, eat a little less.
- Apply this same practice when purchasing fish. In terms of flavor, wild fish is always supreme. However, no one disputes that in the future the vast majority of our fish will be farm raised. Currently, due to the massive consolidation of waste coupled with unnatural feeding practices, fish farming often poses dangers to our waterways and the wild species that

inhabit them. This industry is still so new that it is continually evolving, and a good fishmonger can help differentiate the choices.

- Read labels. In general, the more many-syllable, hard-to-pronounce words, the more processed the food is. If something makes you uncomfortable, don't purchase it.
- Like a good chef, be mindful of the economic bottom line. But don't forget that every dollar you spend is a statement, a small vote for your beliefs. And a dollar spent locally positively impacts the environment on many levels.

The only way that chefs—and everyone else, for that matter—can stay abreast of the ever-shifting landscape is to remain perpetually aware of what is going on around us, constantly seeking information on the ways of the food world and its impacts on environmental issues. Choosing local, organic foods, when possible, not only helps neighboring farmers support their families while maintaining pristine lands for another generation, it also provides the consumer the invaluable opportunity to reduce the damaging fuel consumption incurred by long-distance shipping. Global climate change and our national food habits are not unrelated; by eating responsibly we can each be an integral piece in the stewardship of our planet.

The subjects of how and what the world feeds itself—the customs and century-old rituals passed down from generation to generation, the tremendous influence of science on the evolution of our globe's cuisines, and the environmental impacts of our food choices—are both exhilarating and a bit frightening. And the breadth of choices we have at our disposal is also exhilarating and a bit frightening. The fact that that old chicken processor at the food show could sell me pallets of frozen chicken nuggets shaped like dinosaurs in three different colors but could find no profit in selling me whole, local, farm-raised, antibiotic- and hormone-free three-and-a-half-pound roasting chickens is just a sign of the times.

Sean Kelly is an environmentally conscious chef and restaurateur. He lives in Denver with his wife, psychologist Randi Smith, and his two children, Meredith and Nolan.

COYOTE COMMONS—
BEYOND THE PROVERBIAL LIGHTBULB
By Jackson Perrin and Dev Carey

Since we met in college, more than two decades ago, we've been teaming up to take kids into the wilderness. We've taken vanloads of teenagers to the canyonlands of Utah, the hot springs of Yellowstone, and the mountains of Oregon and Colorado. We've led river trips, backpacking trips, and bike trips.

During each adventure, we tried to teach our students about reverence and risk-taking, responsibility and self-confidence. We took young people out of their modern-day, comfortable lives and plunged them into wilderness, hoping to prove that it was nothing to fear and something worth protecting. We showed that away from possessions and cell phones and potato chips, it was still possible to smile and have fun. We used words like *ecosystem* and *human disturbance*, *trophic levels* and *sustainability*, and as we talked, we let our students see and feel beauty unfiltered by air pollution or noise.

Sometimes it worked: we saw passionate, vibrant young people dedicate themselves to protecting an amazing place for future generations.

But on our trips, we began to notice something else. We noticed that we reached the wilderness by traveling hundreds of miles in vans and pickup trucks. We noticed that even a "no-impact" wilderness trip modeled consumption: we ate food grown in Kansas and packaged in Seattle, cooked with fuel from the Persian Gulf, and zipped ourselves into sleeping bags shipped from Taiwan. We were preserving a few pieces of wilderness, it seemed, at the cost of the global climate.

We also saw that while we might deliver a transformative experience in the wilderness, our students eventually had to return home, where they had to try to maintain their "no-impact" values while managing money, dealing with friendships and romances, and navigating their educations and, eventually, their careers. While some were able to apply their new skills and insights at home, it wasn't easy, and too often, in the complexity of daily life, our lessons faded and were forgotten or just led to guilt and a sense of hypocrisy. Learning how to surf a big hole or climb a peak with a group was fun and rewarding in its own way, we knew, but we longed to teach skills that applied more directly to everyday lives and to the looming environmental challenges confronting all of our students.

So we started looking for ways to reduce the carbon footprint of our teaching—and, while we were at it, bring our lessons in environmental responsibility out of the wilderness and closer to home. And, as with all our teaching adventures, we hoped our students and we would have a lot of fun along the way.

Sustainable living can seem straightforward: buy a hybrid. Install some solar panels. Shop locally. Travel less. And don't forget the organic cotton sheets. All those actions are worthy, but, of course, they're just the beginning. How do you live green without busting your personal budget? How do you design a career that's rewarding, but still allows you the time to live your values by, say, biking to work? How do you find—and live responsibly within—a community that helps, rather than hinders, your efforts to live more lightly? We wanted to get our students thinking about these questions; we wanted their quest for answers to be our next big adventure.

We had some experience with living lightly. When we first moved to the small town of Paonia, Colorado, to teach at an independent community school, we were in our late twenties—a couple of idealistic bachelors. Our tiny paychecks inspired us to make a game out of living on less than $150 a month. We biked everywhere, year-round; traded work on a local elk ranch for our rent; picked apples; and socialized at potlucks instead of restaurants. After a few years, we teamed up with three families to buy a piece of land and built a six-sided house out of scavenged materials for less than $900. (It's still in use today.)

People often assumed we were either miserably uncomfortable or supported by a family trust fund, but the truth was neither: we were supporting ourselves and having a great time doing it. In fact, our lives were a lot like one of our wilderness trips. We helped each other take risks, learned from one another, and enjoyed the satisfaction of reaching our goals together.

With those experiences in mind, we founded the High Desert Center for Sustainable Studies on our land in 2005. Our land had wild places and a great view, but was no rural utopia: it was close to town and lacking in any kind of irrigation water. It was largely made up of rocks, prickly pear, juniper trees, and sage, and a weedy understory that still reflects overgrazing from the late 1800s. We had spent years thinking that these characteristics made our land unsuitable for a school or a camp—after all, didn't we want to spend the summer dangling our toes in a cool mountain stream? On the contrary, as we learned to see beauty in a skunkbrush and spend hot days at the nearest irrigation ditch, it occurred to us that the mundane aridity of our land could be a strength for teaching and modeling sustainability. Our students would be experiencing firsthand the realities of the water-challenged West.

The first group, six game students, came for a monthlong summer course. During our first meetings, we decided that during the program our main challenge would be to keep our food budget below $3 per person per day.

Our students prepared themselves for a month of deprivation, but within a few days they were helping each other come up with creative solutions. They helped small local farmers—chronically short on labor—harvest crops, press cider, and can food, and in exchange they came home with boxes of local vegetables. As word of their work spread within the community, they had more offers from farmers than they could handle, and soon their only trips to the grocery store were for cheese, oil, grains, and some sweet treats once in a while. Their food budget eventually came in well under the $3 limit.

During the month, we also studied sustainable building and helped another young person build a straw-bale cabin with recycled materials and a bare-bones budget. We visited people who were living out their own vision of sustainability, including a town councilman who grows all of his family's food in his small backyard. We also took a backpacking trip—in a wilderness a few miles from our backyard.

We hoped that what the students learned would transfer to their suburban and urban lives at home. They'd learned ways to cook, heat water, build, and garden that cost less money and used less topsoil and less energy, and perhaps these skills might one day help them avoid a big mortgage and free them up financially to pursue satisfying careers. And while their hometowns might not have as many local farmers in need of labor, they'd learned that community connections made it easy to live with less fuel, less money, and less debt.

The next summer, our neighbor Tyler Norris gave us the use of his sixty-acre ranch, a substantial grant of start-up funds, and his expertise in managing nonprofits—a serendipitous turn of events that allowed us to think bigger. For the next two summers, we continued to offer what we thought was a relatively successful curriculum, yet we knew there was more to preparing students to live sustainably. We worried that by spending our days harvesting carrots and drying apricots we were perpetuating a myth—that the only way to live sustainably was to buy land and become an organic farmer. In truth, there are many other sustainable paths, and we began to search for what these paths have in common.

We noticed, for example, that a person's ability to communicate and work cooperatively with others was more important to living sustainably than was their knowledge of biodiesel or composting toilets. We noticed that leaders in business and government are just as central to a sustainable society as organic farmers and green builders. We noticed that success in any sector requires one to (1) articulate a vision, (2) develop a working plan to bring it to fruition, (3) follow it through to completion, and (4) be happy in the process.

So, with help from our first students, we developed a curriculum that does more than teach young people how to use less. Sure, we still recycle our cans, eat local food, and take rainwater showers, but we place even more emphasis on growing the kind of people who have all the skills—both concrete and soft—required to design and pioneer new ways of sustainable living.

Thinking we had finally found the secret to the pedagogy of sustainability, we tried it on our next crop of students, only to be confronted with yet another layer to the mystery. Although our students agreed that these soft skills came before the technical ones, our experience informed us that there was a whole other set of skills even more crucial: those necessary for a successful inner life. Anyone who wants to live a sustainable life has to find ways of being at peace in a broken world, has to leave guilt behind to take action, and has to face—and sometimes accept—the inevitable hypocrisies of daily life. These aren't easy issues for teenagers and young adults to grapple with.

As teachers, we'd thought that our role was to guide our students through a comprehensive project based on sustainability. But this was putting the cart before the horse: students first need lots of one-on-one attention where they can sort out their inner struggles and proceed with a stronger sense of self and the path before them. That, for us, may prove to be the biggest challenge yet.

We've moved from helping kids know simplicity while sitting under an aspen tree at 12,000 feet to helping them know simplicity while immersed in the complexities of shopping and paying the rent. And we have shifted from teaching kids to use a map and compass to orient themselves in the wilderness to helping them use their values to orient their lives. For us, it is the beginning of another adventure that feels not only meaningful and right, but is also a whole lot of fun.

Jackson Perrin is a science educator who enjoys the challenges of living sustainably. He lives with his wife and daughter in their straw-bale house powered by the sun and watered by the rain in Paonia, Colorado.

Dev Carey is a one-man educational think tank who has taught at all levels, from tots to graduate students, in subjects ranging from hitchhiking to botany. More of his writings and projects can be found at www.highdesertcenter.org.

ECO-CONSUMERISM
By Diane Carman

The tradition of the eco-consumer goes back at least as far as my father, Charlie Carman, although in those days he was called a cheapskate. It was a title my dad wore with pride.

Thanks to him, our family was among the last in the neighborhood to own an automatic clothes dryer, the last to have more than one television, the last to have air-conditioning.

One year, we all got electric blankets for Christmas so he could turn down the thermostat even lower at night in our Wisconsin house, where a glass of water left overnight on the floor of my brother's bedroom often froze.

Dad wore his clothes until the pants were shiny and his shoes until they could no longer be resoled, and I never saw a no-deposit, no-return beer bottle in the house until after the old man died.

If anybody had measured his carbon footprint, it would have been a fraction of mine, as I pursue the life of a twenty-first-century carbon-calculating eco-consumer in earnest. While the eco-consumer of my dad's era was defined by what he didn't buy—which was essentially anything that wasn't absolutely necessary—subsequent generations have embraced the philosophy that almost any challenge can be met by hypervigilant shopping.

Now, I may not be as nuts as my friends who insist they won't drink anything but organically grown coffee prepared in the world's only solar-powered coffee roaster—Pueblo, Colorado's, Solar Roast Coffee—but I'll admit I'm a sucker for product with a green pedigree.

So let the record show that I own a Prius, an overpriced swamp cooler that promises to use far less electricity than an air conditioner, and both a messenger bag and long underwear made from recycled plastic pop bottles. I eat only organic bananas because I've been to Costa Rica and I personally want to save the oceans, and the last time I had a serious chocolate craving, I spent an appalling amount of money for a pound of extra-dark, organic, free-trade stuff that, if it wasn't environmentally holy and kind to both the people and the land in Africa where it was grown, it sure as hell should have been.

I read labels and check countries of origin, and even the beer I drink usually comes from local brewers, like New Belgium in Fort Collins, which uses 100 percent wind power and diverts methane from fermentation to generate some of the plant's electricity.

Then again, I have no idea what the carbon footprint is of my favorite New Zealand wine or how much damage I've done to the planet by buying laptops assembled in Southeast Asia—even if they're fabricated from a few recycled materials.

Which brings us to the real challenge: telling the difference between an honest-to-goodness Earth-friendly product and shameless greenwashing. As with all other questions for the ages, the first place I went for answers was the Internet.

Several self-appointed truth squads were there at the ready to cull the authentically green from the imposters on the market. Among them were the Greenwashing Index, headquartered at the University of Oregon, the folks at TreeHugger.com, and Greenpeace's StopGreenwash.org, though the list goes on.

TerraChoice, an internationally respected environmental marketing firm, published "The Six Sins of Greenwashing" to help educate consumers about the perils of the marketplace that await the unenlightened shopper. They are the sin of the hidden trade-off, the sin of no proof, the sin of vagueness, the sin of irrelevance, the sin of fibbing, and the sin of the lesser of two evils.

In a 2007 study, TerraChoice investigated the environmental claims made by marketers of 1,018 products sold in big-box stores and found that all but one made claims that were misleading or just plain false.

At the same time, One Tribe Creative, a Fort Collins branding agency, is working with companies that are committed to sustainability to make sure potential customers know about their genuine environmental agenda—as well as their products. Helping customers differentiate between what's real and what's bogus is the company's mission.

It's no easy job. As a determined eco-consumer, I find myself overthinking nearly every purchase, weighing more factors than I could have imagined just a few years ago.

Recently, I bought an Osprey messenger bag because I liked the product and the company. The fabrics are made of 86 percent recycled materials. The company headquarters are located in Cortez, Colorado, a community that has struggled economically for years. The business conserves water, supports alternative transportation modes, and contributes to environmental nonprofits. Workers even plant trees around town.

A look at the fine print on the website, however, reveals that the bags are made in Ho Chi Minh City, Vietnam, which means they come with a heavy carbon footprint due to the long trip from factory to market.

Reasonable alternatives are hard to find, though. Given the stunning lack of manufacturing capacity in the United States, it's all but impossible to buy a locally built pack. Still, there's no getting around the fact that the messenger bag I use so that I can commute to work by bicycle traveled a long, long way on fossil fuels before I loaded it with my gear to power myself around town on sweat alone.

Just revealing those details is pretty enlightened, though, and some companies are better than others about delivering the hard facts about their products.

Patagonia, a favorite of climbers and outdoorsmen across the country, goes so far as to run the numbers for you for a large sample of its products. It offers Footprint Chronicles on its website, telling customers what they're buying into when they go home with one of their products.

It was there that I learned that my three-year-old long johns are made of 64 percent recycled content, that their production and shipping generated nearly 8 pounds of carbon dioxide, that each one traveled approximately 7,320 miles to get to me, that 10 kilowatt-hours of energy were used to produce them, and that 4 ounces of waste was generated in the process.

It made me feel downright spoiled.

The Patagonia down sweater that I covet is even more of a challenge for the planet. The geese whose down fills the sweater are humanely produced in Gödöllő, Hungary. The recycled fiber comes from Nobeoka, Japan, and the sewing is done in Qingdao, China. That means that each sweater has 20,555 miles on it before it gets to the stores. Each one is responsible for 7 pounds of carbon dioxide released to the atmosphere, 9.4 kilowatt-hours of power used, and 5 ounces of waste generated.

The next time I hear the phrase *global economy*, I'll think about that down sweater.

One percent of the profits from the sale of Patagonia long johns, down sweaters, and other products goes to support environmental organizations around the world, which eases my conscience, at least a little, as does knowing that a whole lot of manufacturers wouldn't think of releasing—or even calculating—information about their products' footprints as long as the profits keep coming.

Probably the one sector of the consumer goods market that is most affected by environmental sensitivity is the food industry. No longer is eating merely a necessity, a cultural totem, or a symbol of social class; now, it's a political statement.

It's hard to find a consumer who doesn't cop an attitude about food, whether it involves an aversion to hormone-tainted milk, factory-raised pigs, mass-produced spinach washed in water contaminated with who knows what, or bottled water shipped from halfway around the world.

For some environmentalists, eating meat is akin to driving a Hummer. One estimate from the Environmental Defense Fund suggests that if every American skipped one meal of chicken a week and ate vegetarian instead, it would equal the same carbon dioxide reduction as if more than 500,000 cars were taken off the roads.

Michael Pollan characterized the growing obsession with the politics of food brilliantly in *The Omnivore's Dilemma: A Natural History of Four Meals*. Surely, he inspired thousands of urban dwellers to look twice at the so-called free-range chickens at the market, to recoil at the ubiquity of high-fructose corn syrup in their diets, and to tear up the Kentucky bluegrass to plant vegetables out back.

Yet, despite the book's stark description of the environmental wreckage caused by industrial farming and the fast-food industry, the year his book was published, McDonald's reported record sales. The company took in $21.6 billion in 2006, and it was not due to a sudden demand for McSalads.

As I take stock of my choices—the messenger bag on the floor by my desk, the organic milk in the refrigerator, the jacket marked *Made in China* hanging on the hook by the door—I realize once again that Charlie the cheapskate had the right idea.

There's no way to live in twenty-first-century America without producing an environmental impact, often a pretty big one. Making careful choices can reduce that, though a person could go mad trying to weigh all the factors that affect the environment with every purchase.

In the end, the only choice that is guaranteed eco-savvy is to travel light in this life by buying only what you need.

As the longshoreman-philosopher Eric Hoffer—one of my father's favorites—once said, "You can never get enough of what you don't need to make you happy."

Diane Carman is director of communications at the School of Public Affairs at the University of Colorado at Denver. She is a former columnist for *The Denver Post*.

PART TWO

NATURAL RESOURCES

Rushing rivers, forested hillsides, red-rock deserts, and snowcapped peaks form the backdrop for our daily lives in the Rocky Mountain West. Scientists believe that even minor changes in temperature affect this landscape significantly—from the type and extent of forestland to the location and volume of snowfall. Mountain pine beetles at both regional and local scales are changing landscapes that once seemed eternal. And while there is no single data link between beetles and regional warming trends, it is widely agreed that beetle outbreaks in past decades were controlled through consistently low winter temperatures not experienced in recent years. This widespread devastation, from New Mexico to British Columbia, is described in a piece originally written for *The New York Times* by Montana-based writer Jim Robbins, and reporter Hillary Rosner looks at the specific impacts of pine beetles in one place: Grand County, Colorado, where they have literally infected every aspect of daily life. Veteran science reporter Michelle Nijhuis reports on the decline of Aspen forests, which is also being attributed to warming trends.

Glacier National Park provides a case study of changes in one iconic landscape. Missoula-based journalist Michael Jamison's essay (and the accompanying US Geological Survey photo series) describes the groundbreaking work of local scientists documenting shifts in Glacier National Park, and the efforts of the National Park Service to interpret the changes and model practices in real time to mitigate them. Internationally renowned geologist Kirk Johnson describes the climate data observed within the five-decade span of his lifetime in the context of the vast geologic time frame he studies every day, and why it's of interest to him as a scientist.

A family train trip from Colorado to Utah leads Salt Lake City journalist John Daley to reflect on changes to the landscape and responses to these changes within his profession.

There is nothing like questioning water supplies to stir up interest in this region. the projections of climate scientists suggest that annual mountain snowpack, which forms the basis for the water supplies on which cities including Los Angeles, Phoenix, Denver, and Las Vegas rely, will decline over the coming years as the planet warms. Included are essays from Marc Waage, Eric Kuhn, and Brad Udall, experts on water supply and demand at city, river basin, and regional levels respectively. And *Outside* magazine contributing editor Peter Heller imagines the impacts of lower flows on some of the West's favorite river pastimes.

Emerging carbon markets are generating new business practices in urban and rural areas alike. Farmers and ranchers, ever adaptive to conditions for survival, have responded to new economic and political opportunities to maintain viability, as described by author Susan Moran.

Developers, in turn, are in many cases becoming more urban, as land-use and transportation planning becomes more closely associated with reduced dependence on automobiles and with the public health benefits of walking and biking. Jocelyn Hittle and Ken Snyder, of the nonprofit organization PlaceMatters, tell how growth patterns are increasingly driven not only by urban service costs, but by a host of environmental health goals associated with denser, new-urbanist patterns that decrease dependence on cars and protect open space and other environmental values. And conservation biologist Tim Sullivan addresses the rising importance of land and habitat conservation in the face of climate uncertainty.

SCIENCE

GLACIER NATIONAL PARK— PORTRAIT OF A PLACE, AN AGENCY, AND A CLIMATE SCIENTIST

By Michael Jamison

High in the alpine wilds of Glacier National Park, where a ragged jumble of Montana mountains reaches unbroken to the Canadian north, a tiny pika crouches beneath an icy lip of bright blue glacier. Both are retreating from the heat, but the summer sun is unrelenting.

"They're small critters," park biologist Steve Gniadek said of the pika. "Related to rabbits, actually."

The pika, he said, would fit in your hand, "although he wouldn't stay there very long."

Question is, how long will pikas—and glaciers, for that matter—stay cradled in the palm of this park?

Pikas and glaciers both live way up high, on steep and stony slopes far above tree line. Neither does well in a warming world, and neither can move fast enough to outrun a rapidly changing climate.

"The pika is becoming a key species," Gniadek said, "because they just might be the canary in the coal mine to warn us about climate change."

Dan Fagre likes that canary-in-a-coal-mine analogy—likes it so much he borrows it to describe the role of Glacier's glaciers.

"This entire ecosystem is dominated by snow and ice," said the US Geological Survey ecologist, "and the snow and ice are dominated by changes in climate."

His is not the stuff of academic obscurity. A full fifth of the terrestrial world is covered by mountains. And a whopping half of all the world's drinking water comes from those high reaches, Fagre said, not to mention water for agriculture, fisheries, and industry. Understanding how those systems are changing, then, is a very serious business, and wild parks are his laboratory.

Fagre's office is just around the corner from Gniadek's, here in northwest Montana's Glacier National Park. It is a flagship park for climate-change research in an agency increasingly attuned to a warming world. The National Park Service (NPS), in recent years, has emerged as an undisputed leader among public land-management agencies in terms of climate-change action and education.

Perhaps that's because all the best science points to a Glacier National Park without its glaciers, a Joshua Tree National Park without Joshua trees, an Everglades National Park without everglades. A Saguaro without saguaros, a Cascade without cascades, a Mesa Verde not so *verde*.

At risk are beaches in the Golden Gate National Recreation Area, ancient petroglyphs in Olympic National Park, forests in Yellowstone National Park, and almost all of historic (and low-lying) Jamestown, Virginia, in Colonial National Historical Park. Imagine Rocky Mountain National Park without its snowcapped peaks, or Isle Royale without its wolves and moose.

"The national parks in general, and Glacier Park in particular, have become the poster child for climate change, and that means they really are stepping up as leaders in both research and education." So said Leigh Welling, climate-change coordinator for the NPS. Prior to taking that post, Welling headed the scientific research and education center in Glacier and before that was a paleo-oceanographer whose job it was to reconstruct past environments.

"What's clear," Welling said, "is we're going to have to make new choices. This is huge and rapid change, faster than we can account for through natural cycles, and it leaves the mountain ecosystems very vulnerable."

In Glacier, as in other parks, these are not predictions. These are happenings.

In 1850, when early European explorers first began documenting what's now Glacier National Park, an estimated 150 glaciers draped the limestone peaks. By the early 1960s, there were 50. By 1998, there were 26, and all were mere remnants of their former size.

In the blink of a geologic eye—150 short years—the park's glaciers shrank from a combined 100 square kilometers to just 19. In the past century, Glacier's year-round snow and ice coverage has shrunk by an estimated 90 percent. And in recent decades, the meltdown has accelerated; by 1998, the park's Jackson Glacier already had receded to a size not expected until 2010.

The retreat speeds as white snow cover gives way to dark ice beneath, and as fracturing glaciers continually present more surface area—just as a smashed ice cube melts faster than one left whole.

It was no coincidence that Al Gore made his climate stand alongside this park's Grinnell Glacier back in 1997. Change here is easy to see.

The trickle-down effects, too, are tremendous. Meadows are creeping uphill, as are trees and other vegetation. Habitats are shifting, water temperatures are increasing, fires are raging. Summertime water flows are slowing as glacial reservoirs dry.

With evidence of worldwide glacial recession and modeled predictions that all of the parks glaciers will melt by 2030, US Geological Survey scientists have begun documenting glacial decline through photography. These photos were taken about eighty years apart. This is Sperry Glacier. Courtesy of USGS/public domain

Spring green-up comes earlier, fall cooldown comes later, wildfire season lasts longer, and snowshoe hares take on winter's white at all the wrong times, exposing themselves to predators.

No part of life's web escapes. Amphibians time their breeding to coincide with snowmelt, and snakes and shrews time their life cycles to feed on baby amphibians. But if the babies come early, then young snakes and shrews don't eat, and then neither do the great birds of prey.

And high on a Montana mountain, pikas and glaciers both recede, unable to migrate from their warming mountain islands.

"We're just beginning to see the tangible manifestations," Welling said, "and it's prompting the parks to step up and lead by example."

An example of that example: in 2003, the NPS launched its Climate Friendly Parks program, with a three-legged emphasis on training park staff about a changing world, helping parks reduce their environmental footprint, and educating visitors about how climate change is affecting national parks. To date, about fifty parks have joined, including Glacier.

National park managers have installed visitor shuttle systems that run on alternative fuels. They've turned to more efficient building practices and renewable-energy sources.

At Zion National Park, thirty propane-powered buses have replaced 5,000 private vehicles per day, eliminating some 13,000 tons of greenhouse gas emissions. And a new visitor center there soaks up sun, with a full third of its energy needs coming from solar panels. That center, with efficient lighting and cooling systems, reduced energy use by about 75 percent, eliminating about 300,000 pounds of greenhouse gas emissions per year.

Across the NPS system, climate teams are looking for new energy solutions, new transportation solutions, new building solutions—and new ways to tell the public about what's changing and why.

On a blustery day in the high tundra saddle of Logan Pass, Glacier National Park ranger Laura Kloeck delivered her afternoon "Goodbye to Glaciers" presentation to a curious crowd of tourists. Researchers such as Fagre have concluded the park's glaciers will be gone by 2030, if not sooner, and Kloeck's audience seemed in disbelief, considering the chill in the air and the ice underfoot.

Along the park's trails are new wayside signs explaining climate change, and new informational handouts, in glossy color like so many alpine flowers, have sprouted in visitor centers.

The "Junior Ranger" booklet has new pages about climate change, and so does the park newsletter. Interested travelers can go to campfire programs about climate change, and to PowerPoint presentations, too.

"But if a park's going to be an educator, and not just the evidence," Welling warned, "then, man, they'd better have their own house in order before they take on that role."

Grinnell Glacier. Courtesy of USGS/public domain

And so Glacier, too, has cleaned house.

"Climate change is the biggest challenge our land-management agencies have ever faced," the National Parks Conservation Association's Will Hammerquist said. "And Glacier has become a real leader. There are some really visionary thinkers in the Park Service, and the hub for that thinking is Glacier Park."

───────────────

The park's historic motor coaches have been revamped to run on propane, and Glacier's entire fleet of vehicles is now on a biodiesel diet. Management bought bikes for employees, for short trips around the headquarters compound, and a new shuttle system has replaced thousands of private vehicles.

Raft guides even talk climate with clients on nearby rivers between the screams of whitewater descent. In addition, the park has joined the DoYourPart! initiative, teaching visitors how to reduce their personal carbon footprints on behalf of their favorite national parks.

In fact, every division in Glacier Park's administrative hierarchy has been touched by a changing business climate.

The interpretive staff has those ranger talks, and the accountants over at Purchasing now shop green. The facility and maintenance crews build and remodel for efficiency, and the scientists gear their research toward monitoring change.

The park has a green team, and staff is carpooling, recycling, reducing the footprint.

It was, of course, a fairly small footprint to begin with, but the new national park role is, after all, to lead by example. When scientists conducted a cursory survey of Glacier Park's greenhouse gas emissions, they found that the 1 million–acre wilderness park produced less than 1 percent of Montana's total emissions.

About 85 percent of park output (some 6,000 metric tons of carbon equivalent) comes in the form of vehicle exhaust. On the other side of that equation, Glacier's sprawling forests sequester an estimated 79,000 tons of carbon equivalent every year, making the park a carbon sink, not a source.

And therein lies the rub. Parks do not cause the problem, but they surely are burdened by the problem and now have been forced to become leaders in finding solutions.

Some folks, according to biologist Steve Gniadek, see climate threats to parks as "a panic situation." But he, for one, sees these changing times as "an opportunity to make a real connection with the public and to educate them about science and how science works."

In Glacier, it works by monitoring pikas—and mountain goats and Clark's nutcrackers and other critters already pushed into the heights. It works by studying glacial retreat—with repeat photography and computer modeling and tree-ring analysis, with dendroglaciology and snow chemistry and geology, with hydrology and forestry, and from the ground and from the air and even from satellites in space.

And here's the thing: no matter how they study it, whether they explore a five-ounce pika or a heavyweight glacier, the story is exactly the same. The emergent narrative is of a warming world and of island parks that cannot move to escape the heat.

"The entire park ecosystem is undergoing an incredibly rapid shift," Gniadek said, "and some species won't fare very well." Those that will fare best, he figures, will be those able to leave the island for better climes.

"These critters move," Gniadek said. "It's critical they be able to cross in and out of the park."

But to get from here to there, he warns, they'll need migration corridors through the ever-increasing tangle of roads and subdivisions and strip malls, the moat of development that rings national park islands.

Another view of Grinnell Glacier. Courtesy of USGS/public domain

The anchors for those corridors, he figures, are the vast expanses of western public lands, a fact that links the future of the nation's parks even more tightly to the future of a changing climate.

Because without intact parks and wild corridors between them, Gniadek said, "some of these species, I believe, with enough time, could just blink out" as climate pressures build. It is, he said, a "whole new world we're working in."

The NPS turns 100 years old in 2016, and its first century, Welling said, was characterized by stability.

"You could count on it," she said. "The mountains and the depth of nature seemed very long-lived. They'd always been here and always would be here."

But now, Welling said, here at the brink of the second century, "we've come to a very interesting turning point."

It turned in 2007, for instance, when Glacier Park's Gem Glacier, for the first time in recorded history, was entirely snow free, a glistening sheet of bare ice sweating dark and blue under a blinding sun. Many miles away, a bubbling mountain stream turned to a trickle, fading finally, underground. It was one of many streambeds that dried up that year, and one of many more to come.

"If the first century was defined by the past and by stability," Welling said, "then the second century is the exact opposite. Now, we're looking at the future and at remarkable instability. Now, all of a sudden, we're faced with how fragile a system like this can be.

"We've always gone to nature because it is so huge and we are so little," Welling said. "We've gone to feel humble, and to put ourselves in perspective. But today, when we go, we're also reminded that we actually affect everything. There's not a place on the planet that we haven't touched. Not even a wilderness like Glacier."

Not a giant glacier, or a tiny pika.

And as for tomorrow, "hold on to your hat," Welling advised, "because by midcentury things are really going to start changing here, in ways we'll all see and feel directly."

Michael Jamison is a print journalist based in northwest Montana. He operates a bureau for *The Missoulian* newspaper, working from the fringes of Glacier National Park and reporting on environmental science issues.

CLIMATE IN GEOLOGIC TIME AND MY LIFETIME
By Kirk Johnson

I first saw Colorado in 1983 when I was driving through on Interstate 70 to attend a lecture in California. I had left Philadelphia thirty-six hours before and was pushing my diesel VW Rabbit hard when a late spring upslope storm dropped a foot of snow on Denver and closed the highway. Against my will, I spent my first night in Denver. The next morning, I wandered around the Denver Museum of Natural History, never even imagining that I would begin my career there eight years later. By midmorning, the pass was clear and I was on the move, and by 3:00 PM the next day, I was sitting in a meeting room at the US Geological Survey facility in Menlo Park as the lecture began. I had only been in Colorado for the time it takes to drive across it, plus the time it takes to eat one steak, visit a museum, and sleep for seven hours, and frankly, I remember very little of what I saw on that quick trip.

The lecture was another thing altogether, and I remember it almost slide for slide. Paleobotanist Leo Hickey had spent the previous summer at latitude 84° north on Ellesmere Island in the Canadian high Arctic and his talk was full of images of the incredible polar landscape and unbelievable 50 million–year–old plant and animal fossils. From crocodile teeth to lotus leaves, these fossils demolished the dogma that the Arctic has always been cold. These ideas were so mind-blowing to me that I approached Hickey after the lecture and told him I would do anything if he would consider taking me along as an assistant on his next expedition.

Twelve months later, I was peering out the foggy window of a DeHavilland Twin Otter as we circled a frozen Stenkul Fiord on southern Ellesmere in search of a suitable landing. After four or five passes and a few touchdowns to test the muddiness of the tundra, the pilot brought us safely to ground, where we unloaded our wooden crates full of food and camping gear. Five minutes later, he was gone and our search for fossils had begun. It was late June, the fiord was still frozen, a herd of musk oxen watched us from a distance, and it felt like we had gone back to the ice age.

Ellesmere is almost beyond description. As large as England but with a population of less than 500, it exists in a truly polar climate. Its southern tip is more than 400 miles north of Alaska's northern tip, and as a result the sun doesn't set for more than three months in the summer, nor does it rise for more than three months in the winter. Here, frozen ground known as permafrost is more than 3,000 feet deep.

Even from our five-tent camp, I could see why we had come. The slopes around us were striped with thick black coal seams. Our first excursions produced abundant fossils of broad-leaved tree leaves and the bones of crocodiles, turtles, and forest-loving mammals. The magnetic orientation of Ellesmere rocks serves as indicators of ancient continental drift and shows that the island has been above the Arctic Circle for the last 50 million years. Similar fossil discoveries from Spitsbergen, Alaska, Arctic Siberia, and Antarctica confirm the observation that I made from my tent on Stenkul Fiord. Simply stated, 50 million years ago, the world was warm enough that there were no polar ice caps and there were forests as far north and south as there was land. Here was something clearly at odds with what I had learned in my geography classes.

Ten years later, I was living in Denver, working as a curator of paleobotany at the Denver Museum of Natural History, and struggling to complete the science and story line of the museum's new exhibit "Prehistoric Journey." The exhibit's goal was to take the history of life on Earth and make it accessible and compelling to the museum's visitors. We used gorgeous fossils and built a series of prehistoric landscape dioramas based on actual fossil sites. It was my job to lead the expeditions that collected the fossils that grounded the dioramas in scientific observation. By 1994, the field trips had been

completed, the exhibit was under construction, and I was working with muralists, taxidermists, plant fabricators, and sculptors to build a series of eight dioramas. I was receiving a crash course in planet Earth, time travelling with a shovel, and literally resurrecting extinct ecosystems. The message of the exhibit was that the Earth has been constantly changing. Evolution and extinction steadily change living organisms while geology and plate tectonics modify oceans, atmosphere, landmasses, and mountains.

So it was likely that I was thinking deep time thoughts when Colorado Department of Transportation paleontologist Steve Wallace walked into my office with a box of fossils that he had found in a road cut on Interstate 25 in Castle Rock, Colorado. People bring me fossils all the time, so that's nothing unusual. What was unusual was what was in the box. It contained six large fossil leaves, each of them a species I had never seen. As a student of the fossil leaves of the American West, it is unusual for me to encounter species that I have not seen before, and here was a box full of paleobotanical novelty. Not only were the species new to science, but they also had several characteristics unique to leaves that grow on trees in modern, wet, tropical rain forests. This was a surprise—a fossil rain forest in Colorado, more than a thousand miles north of the Tropic of Cancer.

My team and I spent the better part of the next decade ripping apart the Castle Rock road cut and collecting more than 10,000 fossils. The site preserves a buried forest floor with intact leaf litter and the rotted-out bases of giant rain forest trees. Like living tropical rain forests, the species diversity of the site is stunningly high. So far, the site has yielded palms, cycads, ferns, and more than 165 species of broad-leaved trees, making it the most diverse and best known tropical rain forest fossil site in the world. Other similar-aged sites in Colorado and Wyoming have produced the fossils of arboreal primates, boid snakes, yard-long philodendron leaves, crocodiles, and other rain forest elements.

As odd as these polar crocodiles and Colorado rain forests seem, they map neatly onto a growing worldwide data set that documents a warm, ice-free greenhouse world that existed between 300 million and 34 million years ago. During this time, sea level was more than 300 feet higher than it is today. About 34 million years ago, our planet transitioned from the warm greenhouse condition to the much cooler icehouse world that we inhabit today. This cooling was caused by the separation of Australia and Antarctica, the northward drift of South America, and the creation of a cold ocean current that circled the newly isolated Antarctica. For the first time in more than a quarter billion years, thick ice sheets began to form on Earth.

This condition persisted for 32 million years, and then things got really cold. In the last 2 million years, the icehouse climate saw more than twenty glacial episodes (ice ages) where thick ice sheets grew south to the latitude of Seattle, Chicago, and Boston. As the ice accumulated and flowed, it grated and ground the landscape, creating the Great Lakes and shaping the topography we see today. These ice ages were separated by shorter warm periods, called interglacials, and we have been enjoying the most recent interglacial for the last 10,000 years. In the long view of all of Earth history, greenhouse conditions are more common than icehouse conditions, but since we are people of the icehouse, it is not immediately obvious to us that we are living in an anomalously cool time.

Our understanding of ancient climates has grown rapidly since scientists began coring the layered sediments at the bottom of the world's oceans and the two-mile-thick ice caps on Greenland and Antarctica. These data sets began to be available in the late 1960s, and they have revolutionized how we understand the workings of the Earth and its oceans and atmosphere. One of the signature findings of the long ice cores of Antarctica is that air temperature and the concentration of carbon dioxide in the air have a close relationship. During periods of extreme ice growth, carbon dioxide is drawn down from the atmosphere and dissolved in the oceans. When the ice melts, sea temperatures warm and carbon dioxide is released back to the air. Carbon dioxide, even in small concentrations, functions as a greenhouse gas, trapping heat in the lower atmosphere and warming the planet.

Like most scientists, I have long been aware of the hypothesis that the burning of fossil fuels in the last two centuries has unnaturally increased the concentration of carbon dioxide in the atmosphere. Still, I didn't make much of the connection between my study of ancient greenhouse ecosystems and modern climate change. There were two main reasons for this. First, my stuff happened naturally, and it happened a long time ago, so it didn't seem directly relevant to something related to recent human activity. Second, it was difficult to measure geologic time with great enough precision to make direct comparisons between deeply prehistoric events and mere historic events. I didn't doubt the idea that recent climate change was driven by humanity's carbon emissions, but I also didn't see how the ancient paleoclimate was directly relevant to the discussion.

This changed for me in 2005 with the flooding of New Orleans in August and the severe reduction of Arctic sea ice in September. The first event showed the potential scale of natural disasters, while the second event made it clear that the polar regions would be where the effects of global warming would first become obvious. I remembered my summers on Ellesmere Island. Looking for insight from the Greenland ice core records, I found the studies of Richard Alley from Penn State that showed that a major warming event 11,600 years ago had happened in less than three years. I began to realize that just because something happened a long time ago does not mean that it took a long time to happen. A second lesson from the ice core data is that an abrupt climate change can have a gradual cause. With this perspective, prehistory becomes an archive of data directly relevant to climate change today.

In November of 2005, I gave a lecture about the relevance of paleoclimate to modern global warming to an audience of nearly 1,000 oil and gas geologists in Calgary. I had innocently assumed that their geologic knowledge would predispose them to understand the significance of climate change. My assumption was not correct. I was startled by the number of attendees who, with their questions, rejected the idea that human activity had anything to do with global warming. Many of them strongly objected to the methodology of using mathematical climate models to predict future climate trends. The pairing of paleoclimate records with climate models creates a smooth framework from the past through the present to the future that allows the creation of testable hypotheses about climate processes. This is solid science with tremendous potential, yet several of the questioners responded to it with dogmatic denial. I had clearly misjudged my audience and it made me want to understand what was going on that would cause such an emotional response to a series of scientific observations and predictions.

In early 2006, Tim Flannery, an Australian paleontologist, wrote a slim and convincing book entitled *The Weather Makers*, in which he clearly summarized recent climate research and the potential dangers of unchecked global warming. In this book, he introduced me to Charles Keeling and his amazing career-long collection of atmospheric carbon dioxide from the sky above Mauna Loa and its rise from 315 parts per million (ppm) in 1957 to 385 ppm in 2005, the year Keeling died. Keeling's story struck a personal chord for me since my parents were married in 1958 and I was born shortly thereafter. Keeling's curve may have been his career, but it is also my life.

When a climatically significant component of the Earth's atmosphere shifts 20 percent in only forty-eight years, it is worth paying attention.

I have increasingly come to realize that Earth's climate past has a lot to teach us about its climate future. Not only do the extremes of ancient climates show what could be possible in the future, but it is also possible to observe the effects of dramatic climate change in the past. Not only can we observe how fast ancient climate changes happened, but we can also see how long it took for the Earth's carbon cycle to return the climate to equilibrium.

One particular event stands out. About 55.8 million years ago and over a period of time of less than 10,000 years, a huge amount of carbon was released into the atmosphere by natural means. The rate of carbon release was huge by normal standards, but significantly lower than the 9 billion tons per year that is presently being released by human activity. The result was a rapid five- to six-degree Celsius increase in temperature that surged tropical ecosystems poleward. The surface soil of this time was so affected by the pulse of warm temperature that it was oxidized to a stunning array of red, orange, and purple. You can see a remnant of this fossil soil by visiting the Calhan Paint Mines, an El Paso County park located thirty-five miles northeast of Colorado Springs. Or, you can go to a Rockies game, since Coors Field is built of bricks quarried from this layer.

While you are watching baseball, you can ponder the significance of this ancient event. The fossil record from land and sea shows that it took about 170,000 years for the Earth's oceans and land surface to absorb the excess carbon dioxide in the atmosphere and return the planet to the temperatures that existed before the event began. For context, realize that the human species, *Homo sapiens*, first evolved about 170,000 years ago. We acquired the benefits of agriculture and civilization less than 8,000 years ago. We started burning significant amounts of fossil fuels less than 300 years ago, and since that time we have unnaturally added more than 400 billion tons of carbon to our atmosphere. We are well on our way to repeating that natural Earth experiment of 55.8 million years ago. I think it is a curious coincidence that our duration as a species to date is roughly the same as the probable duration of our experiment with our atmosphere.

The distant past does appear to be directly relevant to our immediate future. Who would have thought that paleontology not only sees the past, but foretells the future? As a paleontologist, I have always wanted to visit a greenhouse world. In the past five years, I have come to realize that it is a possibility.

Kirk Johnson is the chief curator and vice president for Research and Collections at the Denver Museum of Nature and Science. He is a fossil-leaf specialist best known for his research on the global extinctions that happened 66 million years ago when a six-mile asteroid struck what is now the Yucatán Peninsula of Mexico. He is the author, with artist Ray Troll, of *Cruisin' the Fossil Freeway: An Epoch Tale of an Artist and a Scientist on the Ultimate 5,000-Mile Paleo Road Trip*, which won the Colorado Book Award for best nonfiction in 2007.

SNOW AND WATER

THE RIVER DRY
By Peter Heller

In the summer of 1988, I shared a house with a friend up on Sugarloaf Mountain above Boulder Canyon. John Mattson was a carpenter who could not stop building decks. He built decks instinctively, with flawless engineering, one after another, the way beavers build dams. One deck had a hot tub, one deck was for wine and conversation, one was for sleeping. On another, I used to cool off after he ran the pool table; I'd look up at the constellations and they seemed to roll and click like billiard balls across the black felt of the night. Fortunately, he had a great view: the pine-covered hills shouldering down into the great V of the canyon, and the city and high plains beyond.

That summer, the view became both more disturbing and more beautiful by the day. The West was gripped by severe drought. Forest fires flared everywhere, the worst of which were concentrated in Yellowstone National Park. By late June, the fires there began to close ranks and run together into a terrifying conflagration that would later consume over a third of the park.

Painters love an apocalypse because of the colors. From Mattson's decks, the sunrises and sunsets, filtered through the smoky haze of numerous fires, got deeper, bloodier. The night wind, which I loved to listen to through my open window, rushed through the pines with a dryer and dryer hush, like the parched ghost of falling water. The wind sounded like a spectral creek begging for a drink. It often smelled of smoke.

Falling water. Mattson and I loved it. We were both passionate kayakers. It was how we'd met, a few falls before, on the Gauley River in West Virginia. We'd paddle almost anything, anytime. Everything else we did, we did so we could kayak. He built houses and remodeled; I delivered pizza, taught kayaking.

In Colorado, the season was short and intense, from roughly mid-May to mid-July, depending on the year. Bigger rivers had longer runs of boatable water, but the wild surge that energized our year was short and early. We lived for the pulse and rush of the first big snowmelt.

That year, I remember it didn't last long. A dry winter left the mountains with a meager snowpack. The peaks shrugged off their load in a couple of weeks. A dryer spring meant that the thirsty ground soaked up a lot of the runoff. Mattson and I walked down to our favorite after-work run on Boulder Creek and stared at the slow, clear current as if we might conjure by force of will a flood, tossing waves, fast burbling eddies, foaming holes. The sun beat on the big, smooth boulders of the bank. It was hot. Sweat trickled down my sides. In normal years, the cascade of ice water cooled the whole canyon. Not now.

"What do you wanna do?"

I shrugged. The boats were on my truck. It was sad.

I said, "Do you know how to fish?" I looked at Mattson. He has a wild beard and intense, protruding blue eyes. Forget that idea. His diet is decks and adrenaline. I could feel the malaise of the desiccated country wrapping itself around my limbs like tendrils of smoke.

Not Mattson. He's irrepressible. "Hey," he said, suddenly excited, "let's go up and run upper Boulder Creek. I don't think anyone's ever done it, it's so frigging steep. But we could paddle it now. There's no water in it!"

I stared at my buddy. This is the magic, the defiant logic of an indefatigable optimist.

We did it. We drove up the canyon, along the steep, boulder-choked creek. We got dressed—I wore a neoprene top just to pad my elbows—snugged into our boats, teetered them off a rock, and splashed into a dark sixteen-inch-deep pool. To get over the lip of the first four-foot cascade, we had to grip the paddles in our teeth and push off with both hands. There was some current—enough where it channeled to pin our boats sideways between rocks and drown us if we really tried. So I was on my toes, and I felt with a kind of welcoming exuberance the return of an adrenaline rush. We caught up with each other in the tiered pools and took turns going first over the bigger drops. There wasn't enough water in places to get a full powerful stroke, so some of the moves required launching off a shelf, grabbing a rock with an outflung right arm on the way over, and pivoting. It was challenging. We bruised our arms and shoulders. We got stuck between boulders and had to rock ourselves free. We watched each other like hawks, ready to jump out and free the other should he get seriously pinned. I thought, *This is a parody of kayaking. This is bashing.*

In a short pool above an eight-foot slide we caught our breaths. I said, "Hey, John, is this paddling?"

"I dunno. Got a water bottle?" He drank, grinned at me. His wet beard was stuck down in two mats like a goat's. "It's fun, huh?"

I started to laugh. I couldn't stop. It must have been the relief of being in a boat in the water, even if it was only ten inches deep.

──────────────────────

We make the best of it. I kayak less now, fish more. I have a favorite creek in the West Elks that winds down off the flanks of Mount Gunnison. It pours into a rimrock canyon thick with black timber. A single dirt track strings along it, rising away from the water and leaving the river wild and alone for a few miles in the narrowest, steepest part of the canyon. I put on light waders and walk up the middle of this stretch where I can, kneeing the current and placing my feet carefully on the smooth, algae-covered stones. My feet go numb with cold. The sun is going down upstream and on certain bends it slicks the riffles with molten silver. I step around a boulder and can hear suddenly the threshing of current through a fallen snag, smell the cold, aerated water. The stony banks on both sides are covered in small purple-stemmed willows, sparse grasses, tall stalks of mullein, scratchy hawthorn and alder. White moths flit among purple asters, catching the last warm light. Above the banks rise the dark slopes of spruce and fir. They climb steeply to outcrops of ruddy sandstone.

In a drought year, by late June or early July the creek is already showing its bones. Fallen trees, which usually sift a swift current, lie resting out of water, propped up on dry rocks. Gravel bars split the bends. The rapids channel, exposing wastes of rounded stones. The current slows and warms. The mullein is crumbly dry, the willows slack. I wade in ankle-deep riffles and it's easier and sadder to fish, because the trout are concentrated into the few deep pools and are hungry. They must be stressed by the rising temperature as well. They don't fight with the vigor or confidence of the ice-water evenings when the canyon was pumping. They give up.

I notice more elk and deer tracks in the silt along water's edge—they must be stressed too, having to drop down farther to drink. The wind still stirs downstream after dusk and still carries the pungent scent of the forest, but it is a warmer fragrance, dryer, not fresh, and I miss the scent of cold stone. It's as if the country were gasping for a deluge. In those seasons, I put up my rod earlier and pray most of the fish will make it, and hope for fall rain.

The scientists say that these summers, and ones like the drought of 1988, will become more frequent and more severe. The climate models for the West in the coming decades are not sanguine. I think how the high snowpack is the heart of this country, how it fills and pumps the networks of rivers and creeks that cascade out of the hills and nourish everything. How one doesn't really need models and statistics; it's clear that the warmer winters and earlier springs don't bode well. I can hardly bear the thought of the watersheds drying up.

My friends and I came of age in love with the power and wildness and music of these streams. I still love them. It's a love more prone to terror, more attuned to loss.

Peter Heller is a contributing editor at *National Geographic Adventure*, *Outside*, and *Men's Journal*. He is the author of *The Whale Warriors: The Battle at the Bottom of the World to Save the Planet's Largest Mammals* and *Hell or High Water: Surviving Tibet's Tsangpo River*. His forthcoming book, *Kook: A Memoir*, will be published in the spring of 2010. He lives in Denver.

NO-REGRETS STRATEGIES FOR CLIMATE CHANGE
By Marc Waage

"What's past is prologue."

—From *The Tempest* by William Shakespeare

The Colorado Rocky Mountain region is already warming. The big wild card is whether it will get wetter or drier. A wetter climate would be welcome news for water utilities struggling to meet the water-supply needs of the region's booming population growth, whereas drier weather would bring serious new water-supply problems. Water utilities, challenged with planning for future water needs, are concerned about the uncertainties surrounding climate change. But there is hope. Although climate change presents a variety of threats to water utilities, there are promising new planning methods for reducing those threats.

Our region's water systems have turned our highly variable and often scarce amount of precipitation into a reliable water supply for millions of people, their industries, businesses, and farms, while preserving much of the environmental and recreational amenities that make the area such a great place in which to live. Doing so required water utilities to develop vast networks of water systems throughout the region. Typically, utilities planned these water systems to provide reliable water delivery through the worst drought conditions that had been recorded, going back fifty to 100 years, and usually added a small safety factor to deal with unexpected or changing conditions.

In essence, most water systems in the Colorado Rocky Mountain region were planned with the expectation that weather and supply conditions in the future would not be much different from those experienced in the past. Climate change now is threatening this fundamental planning assumption, and we are a long way from knowing what will happen to our region's water supplies. What types of shortages could be created, and what can we do now to lessen the impacts? How will we continue to provide water for our booming population, and how will we maintain the environmental and recreational amenities of our rivers?

The first challenge in answering these questions comes from Colorado's geography. The state's distance from ocean moisture, its central latitude, and complex storm tracks, combined with the Rockies' influence on local weather, make it one of the most difficult areas to predict how it will be affected by global warming.

We do know the Rocky Mountain region is expected to warm faster than much of the world. The climate models, however, disagree on how much it will warm and the speed at which climate change will occur. Warmer weather brings earlier spring snowmelt and less snowpack for summertime streamflow. Climate scientists expect a warmer atmosphere will cause greater variability in precipitation, making both floods and droughts more likely. They also expect forest fires to increase and predict other ecological changes in our watersheds, including a decline in water quality and an increase in river sedimentation. The pine-beetle epidemic in Colorado's lodgepole pine forests may be a harbinger of these changes.

What is uncertain, though, is whether the region will get wetter or drier. For instance, climate-model projections for thirty to sixty years from now for north-central Colorado are almost equally split between drier and wetter

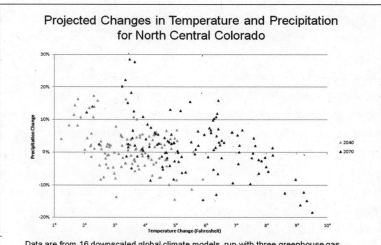

Projected Changes in Temperature and Precipitation for North Central Colorado

Data are from 16 downscaled global climate models, run with three greenhouse gas emission scenarios representing no reduction in growth, a leveling-off by mid-century, and immediate reductions. Results are 30 year averages (around 2040 and 2070) compared with 1950 – 1999. Data courtesy of the Joint Front Range Climate Change Vulnerability Study.

conditions. The wide range of projections in temperature and precipitation is shown in the graph.

We know temperature increases alone could substantially reduce our water supplies. For example, warmer weather causes snowpack and stream water to evaporate faster and forests to consume more water. A simple assessment by Denver Water found that a 5°F warming, which is the median projection for Denver's mountain watersheds in the next fifty years, could reduce Denver's supply by 14 percent—enough water for more than 100,000 households. It would take a significant increase in precipitation to offset these losses from warming.

Not only is the Colorado Rocky Mountain region a hot spot for warming, it continues to be one of the fastest growing areas in the country, particularly in the urban archipelago stretching along the Front Range from Fort Collins to Pueblo. During the past two decades, many water providers in this urban area have focused on developing water conservation, reuse, and smaller supply projects to provide water for future growth. At the same time, water has been dedicated to the environmental and recreational needs of the area, such as endangered species, trout fisheries, and whitewater recreation.

By using water more efficiently and changing the way water systems are operated, water managers have successfully stretched the available supply, or the "water pie," so agricultural, municipal, industrial, recreational, and environmental water needs can be better met. This new water pie has been carefully measured, cut, and apportioned based on the assumption that future supply conditions will be like those of the past. But if climate change shrinks our carefully crafted water pie, water managers will struggle to adapt and divide the pie into even smaller pieces.

Climate models make projections of future temperature and precipitation patterns. To be useful to water managers, these predictions usually are downscaled to local watershed conditions, converted into streamflow projections through a hydrology modeling process, and incorporated into water-system models to determine the effects on local water supply and demand. Fortunately, Boulder, Colorado, is home to many national climate-change experts. Denver Water initiated a partnership with these experts to produce new streamflow projections for Front Range water utilities from Fort Collins to Colorado Springs. The state of Colorado is coordinating a similar effort to estimate possible streamflow changes in the Colorado River in western Colorado. These efforts are important first steps in understanding the potential impacts of climate change on our region's water supply.

Other important implications that must be studied are the effects of warmer weather on water uses, such as lawn watering and agricultural irrigation, and how water might be allocated differently under the state's water rights regulations. Within Colorado, only the City of Boulder, using a federal grant, has worked through the complex modeling process of estimating the potential impacts of climate change on local water supply and use. However, several other Front Range cities, including Denver, have begun similar analyses.

Unfortunately, these analyses start with a wide range of uncertainty from the climate-change models and add more uncertainty in determining possible local effects on water supplies. Some water providers are waiting for more reliable projections before they can justify taking actions to prepare for climate change. This often is called the need for "actionable science," and much of the water industry's response to climate change is focused on the need for improved climate modeling. Because of the enormous complexities of climate modeling, it may be many years before climate scientists make significant breakthroughs in these models.

Another approach is to accept the uncertainty and look for new planning methods that prepare a water utility for more unknowns. Ideally, these new planning methods will identify strategies that prepare water providers for the wide range of possible climate-change impacts. These are referred to as "no-regrets (or low-regrets) planning strategies" because they accommodate many possible future conditions.*

No-regrets strategies are rooted in increasing a water provider's flexibility and diversity while preserving and developing options to deal with changing conditions. No-regrets solutions include maintaining or improving water systems' operational flexibility to adapt to changing conditions, as well as developing a diverse portfolio of resource options.

Conservation and water reuse often are identified as no-regrets strategies to prepare for climate change. While conservation usually is a cost-effective and environmentally friendly way to provide water for growing populations, it may have some drawbacks. When utilities use savings from

---

* The term *no-regrets* has been used to describe actions that would be taken despite the threat of climate change. In this context, the term is expanded to describe robust or resilient strategies that work well in many different future conditions, including the wide range of possible climate-change scenarios.

conservation to serve population growth, they stretch the supply to serve more people, leaving a smaller buffer to adapt to climate change. Reuse of municipal wastewater for nonpotable purposes, such as irrigation and industrial uses, is another favored supply strategy, and several cities along the Front Range have plans to develop most of their reusable supplies. But fully developing reusable supplies may put a water provider at risk if climate change causes the original source of water to shrink and the reuse supply to decrease as well.

Treating water for reuse takes a significant amount of electricity, meaning carbon emissions will increase unless water utilities use solar, wind, or other non-carbon-emitting energy sources. New supply projects have a similar problem because most of the gravity-fed configurations already have been developed, leaving a majority of energy-intensive projects that pump water uphill.

The Water Utility Climate Alliance, a group of eight major US water providers, including Denver Water, is one of the first responders to climate-change challenges. It advocates for more accurate climate-change projections and for new planning methods to identify no-regrets strategies. The group is evaluating several promising planning methods, including scenario planning, robust decision making, decision analytics, and real options. Denver Water is conducting a pilot project on robust decision making, which uses computer modeling to generate many possible future climate-change conditions and to determine if no-regrets solutions can be found. Denver Water also is using scenario planning to address potential climate-change impacts and other major uncertainties in its long-range water planning. Scenario planning is a technique borrowed from the business sector that develops a small number of planning scenarios to represent possible future conditions and identifies strategies that perform well for most, if not all, planning scenarios, at least in the near term. A key goal of scenario planning is to develop near-term actions that prepare a water utility for a variety of future conditions.

Unlike scenario planning, where utilities assume that different future scenarios are equally likely to occur, the decision-analytics method estimates the chances that future events will happen. Then it plans accordingly, saving the utility money in the long run by not investing in scenarios least likely to occur. Real-options planning takes it a step further. This method estimates the chances of certain events occurring and combines it with methods for reducing financial risks. While both of these methods are promising, it may be a long time before climate scientists can accurately calculate the probabilities needed in these two methods.

"What's past is prologue" is a quote from a Shakespeare play about looking to the past to learn about the future. Water utilities have successfully used this method to plan our region's water systems. But times are different now. The uncertainties of climate change are forcing water utilities to rethink the way they plan for future water supplies. In the next few years, it will be essential that water utilities replace their past method of confidently planning for a static climate with a careful, flexible, and adaptable approach.

Marc Waage currently manages Denver Water's long-term water planning. For nearly twenty years, he managed the operation of Denver Water's extensive water-collection system. Waage also worked briefly for the Bureau of Reclamation and the Bureau of Indian Affairs on agricultural irrigation projects. He has a bachelor's degree and a master's degree in civil engineering from Colorado State University and is a professional engineer. One of Waage's favorite activities is recreating in Denver's high-altitude watersheds.

MANAGING THE UNCERTAINTIES
OF THE COLORADO RIVER SYSTEM
By Eric Kuhn

Ever since pioneers first diverted water in 1854 from the Black's Fork of the Green River for irrigation purposes, Colorado River waters were considered available for appropriation and development for beneficial use. Until very recently, the basic assumption has been that if we needed additional water supplies, the water was there. To use it, all that was needed was another dam, diversion structure, pumping plant and canal, or pipeline system. New water rights were perfected through beneficial use of an available public resource.

Today, the focus has changed from one of development to reallocation and risk management. Although a number of projects are still under consideration or being actively permitted in the upper Colorado River basin, there is a building consensus that within the Colorado River system as a whole, the existing demand for water now exceeds the available supply.

The projects in the upper basin being planned today may be developing the unused apportionment of individual upper basin states, but from the system-wide perspective, these projects are reallocating existing supplies. The upper basin's "unused" water is currently in use in the lower basin.

To properly manage a system as complex as the Colorado River basin, the numerous federal, state, local, and private entities charged with managing or using the resources of the Colorado River need a fundamental understanding of the three basic sources of uncertainty they face: hydrology, unresolved legal disputes, and future demands. Addressing these uncertainties will require the adoption of three broad management strategies: identifying and avoiding unacceptable outcomes, maintaining effective working relationships among stakeholders, and an increased integration of science into decision making.

The Basic Assumption Concerning the Law of the River

The term *law of the river* refers to the whole body of international treaties, interstate compacts, Supreme Court decisions and decrees, federal and state laws, and adjudicated water rights that are used to allocate, manage, and distribute the waters of the Colorado River system to its many users. A safe assumption is that the basic tenets of the law of the river will continue to set the boundaries or bookends that will constrain all future management strategies, traditional or new in scope.

Specifically, the obligation of the United States under the 1944 treaty to deliver to the Republic of Mexico 1.5 million acre-feet (maf) per year in most years will continue unchanged. The basic apportionments to the upper and lower basins made in Article III, sections A and B, of the 1922 Colorado River Compact will not be changed. The obligations of the states of the upper division at Lee's Ferry to deliver 16 maf to the upper and lower basins under Article III, sections C and D, of the 1922 Colorado River Compact will remain unchanged. The 1964 US Supreme Court decree in *Arizona v. California* will continue to control the deliveries of water on the main stem of the Colorado River in and below Lake Mead. The 1964 decree, along with the 1928 Boulder Canyon Project Act, the 1956 Colorado River Storage Project Act, the 1968 Colorado River Basin Project Act, and the basic environmental laws, will remain largely unchanged and continue to give the United States, through the secretary of the interior, very broad powers.

Changes to interstate compacts require approval or ratification by each participant state legislature and Congress. Changes to federal laws require either a crisis trigger or supermajorities in both houses of Congress. Within the basin states, water rights that define, prioritize, and quantify the amount of water that can be applied to beneficial use are property rights and, except for abandonment for nonuse, cannot be easily changed, undone, or ignored. Changes in management strategies adopted by cooperative efforts will be allowed and implemented through the existing flexibility and perhaps creative reinterpretation of the existing law of the river.

The Basic Uncertainties

HYDROLOGY

When the 1922 Colorado River Compact was negotiated, the collective wisdom was that the Colorado River system had a total yield of well over 20 maf per year as measured in Yuma, Arizona. In fact, the negotiators believed they were only committing a portion of the available system water. Article III, section F, provided for a future apportionment of the remaining waters.

History has shown that there would be no future apportionment, and, in many if not most years, nature has not even provided enough Colorado River water to cover the original 17.5 maf of water committed for consumptive uses to the upper and lower basins under the 1922 compact and to Mexico under the 1944 treaty.

Using the metric of natural flow at Lee's Ferry, the general rule has been that the longer period of record examined, the lower the estimated mean flow. The 1922 compact negotiators had about twenty years of gauge records. In 1922, the estimated flow of the Colorado River at Lee's Ferry was between 17 and 18 maf per year. At the time the Upper Colorado River Basin Compact was negotiated, in 1948, we had over forty years of gauge data, and the estimated mean natural flow at Lee's Ferry had dropped to 15.7 maf per year. Today, the Bureau of Reclamation's natural-flow estimate, based on the period 1905 to 2004, is about 15.0 maf per year.

A number of well-known studies that have been published using the analysis of tree-ring data have expanded the record back 500 years or more. These paleohydrology studies suggest a mean flow at Lee's Ferry in the range of 13.5 maf to 14.8 maf per year. These reconstructions also suggest that drought periods have occurred that are far more severe and longer lasting than what we've experienced in the post-1905 gauge record.

The prospect of climate-change–induced flow variations adds additional uncertainty. While there is a wide range of results in the different published studies, all suggest a future Colorado River with less streamflow. In a 2007 report, the National Research Council of the National Academies concluded that "the preponderance of scientific evidence suggests that warmer future temperatures will reduce future Colorado River streamflow and water supplies." In late 2008, the Colorado Water Conservation Board issued a synthesis report on climate change specifically targeted toward water managers. This report warns that "climate change will affect Colorado's use and distribution of water. Water managers and planners currently face specific challenges that may be further exacerbated by projected climate changes." The study concludes that "all recent hydrologic projections show a decline in runoff for most of Colorado's rivers."

Given the current demands on Colorado River water resources, even a small change in the mean natural flow at Lee's Ferry will cause serious problems. Among the most optimistic of the published climate-impact studies is the 2006 paper by Niklas Christensen and Dennis Lettenmaier. This study suggests modest reductions in the mean flow at Lee's Ferry in the range of 6 to 10 percent. Most recently, a project by the Western Water Assessment to narrow the results of the various studies suggests the floor for the estimated flow reduction is about 10 percent.

These three credible studies model the current operation of the Colorado River with a sustained 10 percent reduction on natural flow at Lee's Ferry. The recent environmental impact statement (EIS) on the lower basin shortage criteria included an alternative hydrology appendix that used estimated flows at Lee's Ferry published by Connie A. Woodhouse, David M. Meko, and Stephen T. Gray in 2006. The paleohydrology-based trace for the period 1620 to 1674 is illustrative. This period has an estimated mean flow at Lee's Ferry of approximately 13.5 maf per year. The model output shows a number of unacceptable and shocking results. For example, the Central Arizona Project (CAP) would experience forty-seven straight years of shortages, including a number of individual years when the project would divert no water at all. Lake Mead would drop below and stay below the minimum level for the Las Vegas Valley Water District to pump water to its customers (1000' msl) for a period of close to twenty years. California, which has the most senior of the prior perfected rights in the lower basin, would experience occasional large shortages.

In the upper basin, Lake Powell would operate below the minimum storage level necessary to produce hydroelectric power over 60 percent of the fifty-year period, and there were two periods, one of five years and one of twelve years, when Lake Powell would be empty and the upper-division states would have been unable to meet their obligations to the lower basin under the 1922 Colorado River Compact.

The lesson is that without major changes in how we currently manage the Colorado River, even a modest decrease in system streamflows on the order of 10 percent could cause significant unacceptable impacts throughout the basin.

UNRESOLVED LEGAL DISPUTES

It is not hard to understand that with the intense competition for the waters of the Colorado River system and the complex and often conflicting compacts, treaties, and federal and state statutes that make up the law of the river, there are a number of unresolved legal disputes. For the most part, these disputes have been well-known for many decades, but until recently there was little incentive to resolve many of them.

However, since the completion and full utilization of the CAP in the mid-1990s, the secretary of the interior issued interim surplus criteria in 2000 and shortage criteria in 2007. The surplus criteria effort included the resolution of major issues in California, including an agreement that quantifies the individual rights of California's senior irrigation users. This agreement, referred to as the QSA, or Quantification Settlement Agreement, was a necessary prerequisite to the water-transfer agreement between San Diego and the Imperial Irrigation District.

The shortage criteria included a new conjunctive management strategy for the operation of Lake Mead and Lake Powell and the implementation of much-needed efficiency and conservation projects.

Despite the clear progress, important unresolved legal disputes remain to be addressed. The first set involves the Republic of Mexico. The second involves the final remaining unadjudicated Native American water rights within the basin.

Under the 1944 treaty with Mexico, "in the event of extraordinary drought or serious accident to the irrigation system in the United States, thereby making it difficult for the United States to deliver the guaranteed quantity of 1,500,000 acre-feet (1,850,234,000 cubic meters) a year, the water allotted to Mexico...will be reduced in the same proportion as consumptive uses in the United States are reduced." What is an "extraordinary" drought as opposed to an "ordinary" drought? If climate change reduces flows in the Colorado River system, is this a drought or just nature reducing the baseline?

The second Mexico issue is internal to the United States and potentially very divisive. Article III, section C, of the 1922 compact states:

> If, as a matter of international comity, the United States of America shall
> hereafter recognize in the United States of Mexico any right to the use of
> any waters of the Colorado River System, such waters shall be supplied
> first from the waters which are surplus over and above the aggregate of
> the quantities specified in paragraphs (a) and (b); and if such surplus shall
> prove insufficient for this purpose, then, the burden of such deficiency
> shall be equally borne by the Upper Basin and the Lower Basin, and when-
> ever necessary the States of the Upper Division shall deliver at Lee Ferry

water to supply one-half of the deficiency so recognized in addition to that provided in paragraph (d).

Among the unanswered questions are (1) when is there a surplus? (2) when there is a surplus, how is it quantified? (3) where in the basin is the surplus water located? and (4) does the upper basin need to cover transit losses from Lee's Ferry to the Mexican border? The stakes are high for both basins. Is the upper basin's ten-year obligation at Lee's Ferry 75 maf, 82.5 maf, something more, or something in between?

In Colorado, the answer to the upper basin's long-term obligation to Mexico could mean the difference between having enough water or not having enough water to support a large new transmountain diversion and perhaps meeting the needs of a large future oil shale industry. If there is no water for additional Colorado River water for diversion to the Front Range, the only other practical choice may be agricultural conversions in the Platte and Arkansas basins. Not having enough water for oil shale could have similar repercussions for Western Slope agriculture.

In the lower basin, the question concerns the impact on tributaries, primarily the Gila River. In all but very rare wet years, the Gila River system is fully used, and has been for decades. The Gila River has already been the primary driver for several Supreme Court cases. It was the primary reason Arizona refused to ratify the 1922 compact until 1944. And as a practical matter, because of high transit losses through the desert from Phoenix to Yuma, the Gila River can't efficiently make deliveries to Mexico.

Another question is when and how the Mexican treaty delivery-obligation issues will be resolved. Will they be resolved through negotiations or litigation or perhaps the negotiated settlement of litigation? Unlike the 1928 Boulder Canyon Project Act, the 1922 compact does not give the federal government any special status to threaten the states with a secretarial decision.

Up until now, neither basin has had a real incentive to press for a resolution of the Mexican treaty issues, but those days may be ending. The states actually came very close to a showdown in 2005. The current dialogue could force certain issues to the table, and the impacts of climate change may accelerate sustained shortages that cannot be addressed without a resolution of Article III, section C, of the 1922 compact.

Compared with other major western rivers, the Colorado River basin has made progress on Native water rights settlements, but issues remain.

The Navajo Nation covers lands in New Mexico, Arizona, and Utah. The Navajo are in a unique position. The tribe has upper- and lower-basin water interests in New Mexico and Arizona and upper-basin water interests in Utah. The state of New Mexico and the Navajo Nation have reached a settlement covering the Nation's claims to the San Juan River. This settlement has been approved by Congress.

The settlement includes the construction of a water supply pipeline that will pump water from the San Juan River to the Navajo Nation and to the City of Gallup. Gallup is located on a tributary to the Little Colorado River, a lower-basin tributary. The pipeline would also provide much-needed domestic water to tribal users in Arizona. This project raises a number of messy compact issues, such as crediting the upper-basin deliveries for water delivered to Arizona via the pipeline as deliveries at Lee's Ferry. In the fall of 2008, the basin states reached a compromise that allowed the legislation to proceed but reserves a number of tough issues for future battle.

Within Arizona, is there even enough water to satisfy the minimal Navajo claims? Under the 1948 compact, Arizona was apportioned 50,000 acre-feet (af) of upper-basin water annually. A major portion of this water is already in use to supply a large coal-fired power plant outside of Page. What happens if the Navajo claims to upper-basin water, which predate both the 1922 and 1948 compacts, cause Arizona's demands to exceed 50,000 af per year? As a sovereignty, can the Navajo Nation use its water anywhere within its boundaries? Can it deliver water diverted from the San Juan River in Utah to tribal lands in Arizona? For example, as a sovereign, Utah takes the position that it can use its upper-basin water in the Virgin River, a lower-basin tributary. It is seeking federal permits for the construction of a pipeline from Lake Powell to Saint George.

FUTURE DEMANDS

The third set of uncertainties involves the demands for the waters of the Colorado River. This problem is not as simple as it may appear. Planning for and meeting the future water demands in the basin are much more complicated than the traditional demographic-based approaches. Water demands will be impacted both by events in adjacent basins and by realities in the future that will be dramatically different from what we can imagine. The Colorado River is one of four major sources of water for the 20 million people in Southern California's coastal plain (Santa Barbara to San Diego). The other three sources are the California State Water Project, the Owens River aqueduct, and local in-basin sources.

There are significant challenges and uncertainties with each of these supplies. The largest single supply is the State Water Project. This project diverts water from the Sacramento River system in the Central Valley Bay-Delta. From the delta, it is delivered hundreds of miles south to Southern California. The project is facing enormous challenges, including seawater intrusion, Endangered Species Act limitations, environmental restoration, and a lack of system storage. Recent court decisions have limited the water yield available to the project.

The bottom line is that the State Water Project water supplies to Southern California are likely to be smaller in the future. This puts more pressure on the Metropolitan Water District of Southern California (MWD) to firm up its Colorado River supplies. Within California, it has the most junior Colorado River rights. The first step to firming up its Colorado River supplies was to transfer existing senior agricultural uses, which it has already done with some success. Its aqueduct has a capacity to pump 1.2 maf per year. Through agricultural transfer fallowing and conservation programs with Palo Verde and the Imperial Irrigation District (IID), MWD has increased its Colorado River diversions. However, will politics in the IID allow more transfers? If not, where will MWD turn? Will its efforts ultimately lead to the upper basin?

Likewise, central Arizona has three major sources of supply: the CAP, the Gila/Salt River system, and groundwater. Groundwater is already over-tapped and aggressively managed. The CAP is the most junior project in the lower basin and potentially subject to prolonged periods of shortage. The Gila River system, including its major tributaries, the Salt and Verde rivers, is a vital supply that has historically provided approximately 1.5 to 2.0 maf of water per year for irrigation and municipal purposes. The Salt/Verde system drains the Mogollon Rim and the White Mountains. Compared with the Colorado Rockies, this watershed is at a low elevation, 7,000 to 10,000 feet. The current climate science suggests that the southwestern United States and lower-elevation watersheds will be the most susceptible to climate change.

Thus, Arizona faces a future when its local supplies are reduced by climate change, its CAP is subject to prolonged shortage, and its groundwater basins are already overtapped. What are Arizona's options? Are strategies such as the construction of large desalination facilities in Mexico on the shores of Baja California politically or economically feasible? Strategies such as aggressive reuse, the desalinization of local brackish groundwater, and the lease of senior Native American agricultural rights from the Arizona side of the main stem appear more likely. At the 2008 convention of the Colorado River Water Users Association in Las Vegas, a water planner from the CAP suggested that in the future, Arizona might build a pipeline from the Mississippi River.

In the upper basin, the major demand uncertainty is energy, specifically oil shale development. With the recent cost of oil and with geopolitical concerns, there has been a surge of interest, primarily at the political level, in developing oil shale. The development of oil shale may require the consumptive use of large amounts of water for oil shale processing, reclamation, necessary electrical power generation, and the associated municipal use by the supporting communities.

The bottom line is that a large oil shale industry (greater than 1 million barrels per day) could require the use of all of Colorado's remaining unused Colorado River Compact entitlement, perhaps more. If the ultimate oil shale extraction technology is new and different from what is currently under development, the resulting water demands could be smaller.

This presents Colorado with a difficult policy challenge. Does it reserve a major portion of our unused water (if we have any) for a future oil shale industry? If not, what are the consequences of the industry turning to the market (agriculture) to meet its future supply needs? The situation is complicated because the energy companies already hold valid conditional water rights (rights not yet perfected by use). If the industry develops its relatively senior rights, the result could be an unacceptable reduction in the yield of existing perfected water systems, including many transmountain diversions.

Three Strategies to Help Manage Uncertainty

To help manage these uncertainties, the basin states have adopted three broad strategies:

1. The early identification, acceptance, and prioritization of unacceptable outcomes. The compilation of a list of unacceptable outcomes is probably very easy; the problems and challenges are reaching a consensus on prioritizing the list and identifying a plan to meet priority needs.

Within the basin, we all know that there are events we accept as model output, but we really understand they will never happen. For example, would a future secretary of the interior ever let Lake Mead drop below the minimum level necessary to deliver water to Las Vegas? The answer is almost certainly no. However, unless Arizona, California, the upper basin, and the other parties get something they want in return, will they publicly acknowledge this reality? What if the cost to the other uses and resources of meeting this top priority is considered unacceptably high? What happens if the owners of the most senior rights or resource agencies say, "No more"?

2. Maintaining positive relationships among the stakeholders. Again, this task is probably easier said than done. In the upper basin, the 1948 compact created an Upper Basin Commission. This commission has served a bonus role of fostering good relations and effective communications among the upper-basin states. However, no similar organization exists in the lower basin or the basin as a whole.

In recent years, the states have done reasonably well at working out consensus solutions, but they have been criticized for excluding other stakeholders. Additionally, the motivation has most often been the threat of a unilateral decision by the secretary. The future challenges may overwhelm voluntary cooperation among the states. The United States Supreme Court has provided a dispute resolution forum. However, using the courts for dispute resolution is both expensive and time consuming. The 1964 *Arizona v. California* decision took over a decade to resolve. The recent Arkansas River dispute between Colorado and Kansas was almost two decades long. Finally, courts can make decisions and interpret laws and compacts, but they cannot provide practical and long-lasting solutions. At the end of any future litigation on the Colorado River, the parties would still have to work out cooperative and practical solutions.

3. Better integration of science into decision making. Again, this is a goal that can be readily agreed to by most stakeholders. The real challenge is implementation.

In recent years, there has been progress. For example, Reclamation's shortage criteria EIS included a nontraditional hydrology appendix. The analysis examined how the system would operate based on the long-term reconstructed gauge record at Lee's Ferry and stochastic hydrology techniques.

Colorado is aggressively pursuing new science-based studies. A number of major water providers are conducting a Front Range climate-change vulnerability assessment. The Colorado Water Conservation Board is conducting a Colorado River water-supply-availability study that will look at vegetation changes, paleohydrology, and climate change.

Many agencies are now assuming that the future will look like the past. They are planning for a number of reasonably foreseeable alternate futures. Reasonable futures include a Colorado River with reduced streamflows from climate change, a future with a significant oil shale industry, a future in which there is a huge worldwide demand for US agriculture, a future when public health requires ultrapure drinking water, and a future with one or all of the above. Can we develop a strategy that does not result in unacceptable outcomes under any of the possible futures? The future includes ecosystem management, fisheries, and wildland fire strategies, not just the traditional water systems for human purposes.

The collection and analysis of basic data will be fundamental to the understanding of the Colorado River system and future management decisions. Without basic data, how will managers and planners know the baseline or understand the effects of climate change? There is no substitute for better science and basic data.

Eric Kuhn is the general manager of the Colorado River Water Conservation District (River District). The River District is the largest and oldest of Colorado's four conservation districts. It was chartered by the Colorado General Assembly in 1937 to "preserve and conserve for Colorado, its Colorado River compact entitlement." The district covers the Colorado River basin except for the San Juan and lower Dolores river basins. Kuhn began his employment with the district in 1981 and became manager in 1996.

WATER IN THE ROCKIES—
A TWENTY-FIRST-CENTURY ZERO-SUM GAME
By Brad Udall

We are already seeing the effects of climate change in Colorado and around the West. Temperatures have warmed by over 2°F since 1970. Spring runoff is occurring earlier in almost all snowmelt basins in the West. A greater proportion of our annual precipitation is now coming as rain instead of snow, even at our highest elevations. Forest fires in the West since 1986 are significantly bigger, longer, and more destructive, and these changes highly correlate to warmer temperatures. Droughts are more severe and last longer. The recent mountain pine beetle epidemic—caused partly by climate change, partly by natural cycle, and partly by human fire management—is now at 2 million acres and is fundamentally changing our mountain landscapes and mountain hydrology. Recent state-of-the-art studies have attributed many of these western effects to warming caused by greenhouse gases.

All of these impacts have a strong connection to water. In fact, changes in water availability, not higher temperatures, will be the delivery mechanism for many of the most significant impacts of climate change. Additional heat will fundamentally alter the water cycle—the vast solar-powered cycle that evaporates huge quantities of water from the oceans and moves that water to land every day. The water cycle, the primary mechanism for redistributing heat on the planet, moves heat from places where there is too much, such as at the equator, to places where there is too little, such as at the poles. Big ocean currents, such as the Gulf Stream, and water vapor carried in storms are two critical mechanisms used by the Earth to transport heat poleward. These very large movements of heat determine our weather. With additional heat due to climate change, we will experience significant changes in the patterns of weather and water in the twenty-first century, the very definition of climate change. The western United States will experience the brunt of these changes.

Climate Change Will Be Serious

The projections for Colorado in 2050 are for additional warming of 2°F or 2.5°F, and, depending on greenhouse gas emissions, the West could warm up by as much as 9°F by 2100. This is an enormous amount of warming; for comparison, the last ice age, which featured mile-high ice sheets over Chicago, was only nine degrees cooler than the average temperature in the twentieth century. A variety of impacts will result. Shorter winters will occur with earlier snowmelt runoff—in some cases, up to sixty days earlier. Summer flows in our rivers will be lower and water temperatures higher, presenting a challenge for all aquatic life. Higher water temperatures speed up biological activity, yet less dissolved oxygen will be available. Trout and other species will suffer. Evaporation will increase, producing more drying of soils, and individual storms will produce more, and more intense, precipitation. Paradoxically, we also expect the number of dry days between storms to increase. Current models indicate that annual precipitation will change little in Colorado, but the seasonal pattern will change—more in the winter and less in the summer. Most climate models predict storm tracks will move northward, further drying out the American Southwest.

Population and Demand Growth as Additional Stress

Climate change is not the only serious problem facing water managers. Colorado growth rates, some of the highest in the nation, also present special problems. In the 1990s, the state grew from 3 million to 4 million inhabitants, a rate of 2 percent per year, and by another 1 million from 2000 to 2008. The projected population for 2020 is 6.3 million and for 2035, almost 8 million. Most of this growth has been, and will continue to be, on the Front Range, where the water is not. With 8 million people in 2035, Colorado will have to supply double the water needed by towns and cities in 2000.

No one seriously doubts that this growth will abate anytime soon. Hotter temperatures in Arizona, Las Vegas, and even Southern California, some of the most desirable places to live in the late twentieth century, will make them far less appealing. Colorado's relative coolness in the future—it is the highest state in the nation—is likely to make it an even more desirable place to live as climate change unfolds.

New Water Supplies: A Zero-Sum Game

Supplying water to all of these new migrants would be a challenge even without climate change because water in the twenty-first century is already a zero-sum game, with winners and losers for every transaction. All water in Colorado is already in use, and any new use is just a reallocation from one use to another, the very definition of *zero-sum*. (In fact, water has always been a zero-sum game, except that for the last hundred years the loser has always been the environment, a stakeholder without a vocal constituency until 1970.) There are currently only four options for obtaining new water supplies in Colorado: taking more from the Colorado River, transferring water from agriculture, municipal recycling of water, and conserving water. All of these options, unfortunately, have serious limits.

In order to understand the four potential sources, the two kinds of water use need explanation. Consumptive uses involve some form of evaporation or use by plants, *transpiration*. In both cases, water is lost to the atmosphere as water vapor. Agriculture and watering lawns and trees are examples of the consumptive use of water. Nonconsumptive uses result in dirty but undiminished return flows that appear at a wastewater treatment plant. Indoor uses such as showers and toilets are nonconsumptive. Agriculture also has a nonconsumptive use component: flood irrigation generates return flows to streams, either occurring on the surface or as shallow groundwater. Return flows, regardless of quality, are highly valuable and are now fully used in Colorado. Most water users, both agricultural and municipal, depend on return flows from upstream wastewater treatment plants and from excess agricultural irrigation.

THE COLORADO RIVER QUANDARY

The Colorado River is by far the largest river in the state, and the only one that *may* have any "extra" water. Extra water is defined as water that is legally available for diversion. This water is not required by compacts, court decrees, or environmental laws to remain in the stream; however, this water may not be extra to a whitewater rafter, a trout fisher, the environment, or a junior diverter who currently relies on it. Through a number of transmountain tunnels under the Continental Divide, the Colorado River provides nearly half the water used along the Front Range. Denver, northern Colorado, and Colorado Springs, as well as many smaller Front Range cities, depend on the river for a significant portion of their needs because our native South Platte supplies are not enough to support the large Front Range population.

Unfortunately, multiple studies over the past thirty years—including nearly ten in the last five—suggest that Colorado River flow will decline 5 to 20 percent by 2050, with greater reductions possible by 2100. With 20 percent less water in the river, the issue is not how much water is left to develop, but what existing water uses will be curtailed so that the state can meet the delivery requirements of the 1922 Colorado River Compact. That agreement, now known to be based on faulty assumptions, requires that the four upper-basin states deliver from Lake Powell a set amount of water to Arizona, California, and Nevada. If the upper-basin states can't meet this requirement, they are legally required to shut down all post-compact water rights, meaning a large portion of the municipal supply used along the Front Range. At least that's the theory. In practice, what will happen is that these municipalities will pay pre-compact Western Slope agricultural interests significant funds for the temporary right to use their water during this period.

The fundamental problem with using the Colorado River as a source of new water is that any new use puts existing senior water users at risk. Every new Colorado River diversion, no matter how small, increases the long-term risk that Lake Powell will not have the necessary water to make required compact deliveries. This zero-sum game involves a hidden and undesirable water transfer from senior users to the new diverter through time: in essence, the new diverter could use water for some years until the compact required all diverters to stop, not just the party who caused the overuse.

Agricultural Water Transfers

The second way to obtain new water supplies is to transfer it from agriculture, which uses roughly 80 percent of all water in this state. Long before our cities existed, agriculture was using great quantities of water, and because it was first, agriculture in general is associated with very senior water (property) rights. The economic and societal problem with this much agricultural use is that most of the water goes to very low–valued crops like hay. In a perfectly optimized economic system, water use would cascade from highly valued uses in cities and industries to lower-valued uses like most agriculture, with shortages being delivered in low-water years to the lowest-valued uses first. Alas, we don't live in an optimized world, and agriculture generally is legally entitled to the first call on our water supplies.

While transfers from agriculture to our growing cities seem to make sense, there are significant equity and scale issues. Large water transfers from agriculture threaten the existence of small towns—take away enough farmers and the entire economic fabric of a community can unravel. Agriculture is a proud part of our history, and many people now believe that locally grown food is a desirable goal. Temporary, rather than permanent, transfers from agriculture can provide our cities with a reliable supply during drought, and this mechanism also has the benefit of meeting the economic ideal of cascading uses. Such temporary transfers can provide farmers with a significant source of income for not growing crops. Some of these temporary-transfer agreements already exist, and more are no doubt being contemplated. Permanent transfers, in many cases highly controversial, have been heavily used in the past and will likely also be used in the future.

Conventional wisdom holds that agriculture is tremendously inefficient and that significant water savings can be obtained by water conservation. While there may be some opportunities for agricultural conservation, the truth is that wasteful agricultural water use is rare; what appears to be wasteful use actually runs to the stream as return flows. For example, Colorado State University scientists have found that large agricultural return flows from flood irrigation in a system like the South Platte make the system operate quite efficiently. Return flows from spring-water applications occur in the fall, keeping more water in our streams during low-flow season. Other recent studies have indicated that agricultural water conservation, such as drip irrigation, may actually increase water use. In general, the supposed savings from agricultural water conservation are overstated. Agricultural water transfers and water conservation are also zero-sum games—less water applied to crops translates to either fewer crops or fewer return flows.

WATER RECYCLING

The third way to obtain new water is to use technology to make clean water from dirty water. Wastewater treatment plants already do this, but technology has advanced to the point that we can now safely drink the treated water. The public currently finds this concept unappealing, but it is very safe and can produce higher-quality water than nature. The best example of this technology is Aurora's new $500 million Prairie Waters Project, which treats wastewater to a very high standard, injects it into the ground, and then extracts it a few months later for use. In essence, using energy and technology, we clean up water faster than nature and then reuse it.

Unfortunately, if recycled water is reserved for consumptive use, then wastewater treatment return flows are reduced and the downstream entity currently using these flows will be harmed. Colorado water law allows and even encourages cities with transmountain water rights to reserve such recycled water for consumptive uses because this water is not native to the river system in which it is being used. Though true, all return flows in the state are now fully used and recycling will deprive a downstream user, or the environment, of such water. Through this lens, recycling really only creates "new" water for nonconsumptive uses. Thus, yet another form of zero-sum game exists for recycled water.

MUNICIPAL CONSERVATION

The final option for new supplies involves municipal conservation. Cities can conserve either indoor nonconsumptive uses or outdoor consumptive uses. Each form of conservation has important nuances.

Indoor conservation keeps municipal reservoirs higher but reduces return flows. Water use accounts for a surprising amount of energy use—treating, heating, and pumping are significant—and hence indoor conservation also has the added benefit of reducing carbon footprints. Ideally, indoor conservation would be for more growth-related indoor use, resulting in no changes to return flows. Indoor use accounts for about half of all municipal water use and changes little from month to month. It is relatively inflexible and much more difficult than outdoor use to curtail during times of drought. Indoor conservation is the one exception to the zero-sum game involving new water supplies: in most cases it has no drawbacks.

Outdoor water use represents the other half of municipal water use. Outdoor water conservation keeps municipal reservoirs higher, and has no effect on return flows because it is usually fully consumed, but it can result in less greenery. Outdoor watering helps to make our cities livable, and watered, green surfaces—be they lawns or trees—are significantly cooler than unwatered surfaces and help counteract the urban "heat island" effect. Water managers rightly complain that permanent outdoor water conservation—such as turf removal—that is then used for new growth, especially inflexible indoor use, removes some of the useful operating slack from their systems. They refer to this as "demand hardening." Like indoor conservation, outdoor conservation is generally laudable, but too much can make our cities less livable and can reduce operating flexibility during droughts.

Two More Collisions Ahead

Aside from the zero-sum games involved in any new water use in Colorado, twenty-first-century growth and climate change are on a collision course with nineteenth-century water law and twentieth-century water infrastructure management.

Our nineteenth-century water law, an artifact from mining camps, is ill-suited to today's world. As a society, we can't afford to have senior low-valued agricultural rights trump our economic powerhouses, our cities. In a serious drought, does anyone really think that an agricultural right, no matter how senior, should be able to continue to divert water when the City of Aurora, with 300,000 people and junior water rights, has no water? The real problem is that our legal system provides little guidance as to how these situations will be resolved, but resolved they will be; one does not fallow a city under any circumstances. The general consensus is that the solution in a serious drought would involve cash payments from our cities to agricultural water rights holders. But at what price, when, and exactly how are unknown. The current legal structure provides far less certainty than there should be; it would be far better to have such sharing arrangements agreed upon well before they are needed.

The other major collision involves our twentieth-century water infrastructure and what climate scientists call "climate stationarity." The primary problem with climate change for water managers is that all water planning has been based on the idea that supply and demand would fluctuate within known historical levels due to a stable but annually varying climate. Unfortunately, our climate is changing rapidly, and the water cycle will especially change in fundamental ways in the twenty-first century. This means the yield of our entire water infrastructure, which includes over 150 major reservoirs and 6 million acre-feet of storage, is much less certain and may be lower than historical records would suggest. Climate change may convert our zero-sum game to a negative-sum game.

A traditional view is that we will just build more reservoirs to handle growth and climate change. However, due to environmental considerations, cost, lack of good sites, interstate compact requirements, and even the carbon footprint of new projects, our reservoir-building options are very limited. Yes, Colorado will likely build a few reservoirs in the early part of the twenty-first century, but these new structures will do little to change our overall infrastructure and will not fundamentally alter the overall water-management problem brought on by climate change and growth. Those who think that we'll just build our way out of this predicament don't understand how and why water management has changed so fundamentally in the twenty-first century.

Concluding Thoughts

Almost all of our new supply options for handling growth and climate change will impact existing water uses. None of the four options is a panacea, and most are fraught with problems, some occurring now and some later. Taking more Colorado River unfairly transfers significant risk from new diverters to current users. Water recycling is great in theory, but requires energy and can't generate new water for consumptive uses without impacting existing downstream users. Agricultural transfers, preferably temporary and accomplished with consideration of equity and social concerns, may provide some water without overly impacting an important activity. But too many such transfers will put agriculture at risk. Municipal water conservation works well for energy conservation and is generally laudable, but can make these systems more vulnerable to drought through demand hardening.

The intersection of climate change with our limited water supply is just the first of many problems with building a world that is truly sustainable. In the twenty-first century, we have already entered a new era of limits on water usage. Solving our water problems will ultimately require clear understanding of the zero-sum nature of every water problem and a willingness to make decisions based on what is "less bad" rather than what is "best."

Brad Udall is a research scientist and the director of the National Oceanic and Atmospheric Administration–funded Western Water Assessment at the University of Colorado. He studies the impacts of climate change on the Colorado River and the West.

LAND

ZEPHYR TO ZION—
TRAIN OF THOUGHTS IN A WARMING WEST
By John Daley

"Momma, Dadda! Look, an eagle! A bald eagle!"

Julia leaps from her seat, flies to the window, smushing her face against it, tapping her index finger on the cool glass. If there's any more perfect sound than your kid giggling and shrieking with wonder at her first sight of this magnificent raptor, I don't know what it is. For Julia, it's pure joy. For me, the excitement is tempered by serious anxiety.

Two days after Christmas, we're taking the California Zephyr from Denver, where I grew up, to Salt Lake City, where we live now. It's a fourteen-hour journey, climbing up from the plains, traversing the foothills into Colorado's high country, crossing Utah's lovely, lonely eastern deserts, then up and over the spine of the Wasatch Range to Provo and into the Salt Lake Valley.

Our sleeper car is a cozy traveling cocoon, comfortable and warm. We watch the world go by. We read, gorge robotically on far too many munchies, and listen to tunes from the old laptop: some Dylan, a shot of Neil Young, a splash of Radiohead, and a smattering of Earth, Wind & Fire. As the train bounces along, the high-pitched screech of steel wheels grabbing steel rails, the acrid smell of diesel combustion drifting back from the locomotives, I have to grin. The next item on the shuffle: "Train in Vain" by The Clash.

Making the trip by Amtrak is definitely not the most efficient way to travel. You can drive between the two cities in eight and a half to ten hours, depending on how badly you want to risk meeting Mr. State Trooper. You can fly it in less than an hour. But by rail, despite occasionally getting parked on the tracks while waiting for the endless parade of coal trains to pass, you are confronted with the sheer vastness of the West. You cannot tune it out.

Glenwood Canyon is the sweet spot of the entire journey, and this year it's spectacular. The rugged gorge is covered in a deep layer of fresh snow. The Colorado River gently slides by, in places slipping ghostlike beneath thin, luminous sheets of ice. Sheer walls of Cambrian rock and limestone rise a thousand feet above the Colorado River. The sun is a hazy platter glowing low in the December sky behind whisper-thin gray clouds.

The scene is already gorgeous enough, and then we start to spot the wildlife. First, it's a bald eagle, a male in full plumage. He's sitting, like a statue, on a telephone pole not twenty feet away. Later, we spot three more baldies, including one gliding just above the river. Ducks and geese float by in the black water. The occasional deer family wanders down for a sip. Farther down the canyon, a small herd of elk munches on grass sticking up through the snow.

Julia shouts, "Whoa! What is that?" Bighorn sheep, we explain. There are five or six of them, all with dinky, eight-inch, scythelike half horns, starting to develop that distinctive curve. "Momma, did you see that? Cool!"

Julia's enthusiasm is a nice distraction from my angst, which was triggered a few miles back when I noticed a stand of copper-colored trees in the middle of a forest of evergreens. Soon, there were more. An entire hillside of what formerly was green is now bleeding red, completely dead thanks to the pine-beetle infestation, which is steadily drilling through vast swaths of forest throughout the region.

Warming: Danger Ahead

In a warming world, it's an ominous harbinger of things to come. A recent Massachusetts Institute of Technology study found that if our current rates of carbon emission remain unchanged, the planet could see a temperature hike of nearly 10°F, a staggering jump. What chance is there, then, for these iconic western pine forests and all that awe-inspiring wildlife?

Like my parents', some of my fondest memories are of skiing and climbing in these mountains, tubing and rafting icy mountain rivers. What will this place be like when Julia is an adult? And what will it be like for her children?

As the train rumbles into the dark canyon shade, the layer-cake walls towering high above now, I wonder: how many of the hundreds of people on this train really understand the changes we're seeing right out our window and what may be responsible for them? If they're like many Americans, they're actually increasingly skeptical about the reality of global warming, according to the latest Gallup poll (2009). It found that 41 percent of those polled believe the threat of climate change is exaggerated.

That's an amazing statistic when you consider this: many of the scientists, the people who've been spending years now documenting what must be one of the most thoroughly studied phenomena in history, are more worried than ever. One study released this year by the National Oceanic and Atmospheric Administration predicted dust bowl–like conditions in the Southwest and elsewhere, which would be "largely irreversible" for a thousand years if we continue burning carbon at our current pace. Another, by the British Scientific Committee on Antarctic Research, declared that Antarctic glaciers are melting faster and across a larger area than previously believed. The head of the United Nations' Intergovernmental Panel on Climate Change says that the current trajectory of climate change is now much worse than originally projected.

How can it be that public opinion lags so far behind reality? Here's my perspective: the nosedive of the journalism world is directly undermining the quality and quantity of climate-related information US citizens consume. Consider the findings of the 2009 State of the News Media report from the Pew Research Center's Project for Excellence in Journalism. The report found that the journalism world, like the financial sector, is in a meltdown. The global economic downturn is hammering the very advertisers that fund most news outlets.

No one is immune: newspapers, magazines, local and national television, radio, ethnic, and alternative news outlets are all feeling the pain. Some are struggling to survive; for others it has proven fatal. A case in point is the paper I grew up reading. Just short of its 150th anniversary, in February 2009 the *Rocky Mountain News* permanently stopped its presses.

The danger is clear, as *The Nation* points out: "As journalists are laid off and newspapers cut back or shut down, whole sectors of our civic life go dark." Sectors like coverage of global climate disruption.

With the news world in the fight of its life, comprehensive, in-depth, nuanced reporting by experienced journalists on such a complex, multifaceted story as global warming is falling by the wayside. Just last year, CNN cut its entire science, technology, and environment team. The network insisted the move wasn't tied to the economic downturn. But one has to assume it was a simple business decision. With ad revenues shrinking, something had to give. In an election year during a severe recession, it wasn't going to be the network's political or business reporters getting the ax.

Still, it's a matter of priorities—some would argue, badly misplaced priorities. Stanford climate researcher and policy analyst Stephen Schneider recently hit the nail on the head when he sharply criticized that CNN move: "You are screwing up your responsibility by firing science and environment reporters, who are frankly the only ones competent" to cover the climate crisis.

While the news from the front lines of that crisis gets steadily worse, you wouldn't know it from coverage in the mainstream media. For example, the Tyndall Report, which monitors the weekday newscasts of the three American broadcast networks (ABC, CBS, and NBC), ranked global warming/climate change eighteenth in the top twenty stories of 2007. In 2008, it didn't even make the list. Within the environmental coverage category, global warming in 2008 was deemed less newsworthy than stories about recycling and plastic bottles leaching synthetic estrogen hormones.

In the twenty years Tyndall has been tracking television news coverage, during what years, in terms of overall minutes, was environmental coverage at its peak? Was it the Gore-inspired 2006? Nope. It was during the late eighties and early nineties, when the scientific community first started sounding the climate alarm.

Likewise, the Pew "State of the News Media" report found in 2008 that the news agenda of the mainstream media was dominated by the presidential election and the economic crisis, with an occasional "one week wonder." The study notes, "Even when a major story managed to break through the clutter of the election and economic coverage in 2008, the press quickly seemed to tire of it." Like comets, stories about the violent conflict between Russia and Georgia, or the sexual indiscretions of Eliot Spitzer, flashed upon the scene and soon faded.

Under current conditions in the journalism world, one could easily envision the following scenario: a huge chunk of Greenland crashes into the sea, the story gets blanket coverage for a week or two, and then it's on to something else.

It's all baffling, really. "Don't bury the lead" is one of the cardinal rules of journalism. But when the story broke about that study projecting a nearly 10°F spike in temperatures this century, was it front-page news or leading newscasts? No. We're talking about temperatures not seen on Earth for eons, a period of time measured by those strata in Glenwood Canyon. Still, apparently it's not the kind of story that deserves the "flood the zone" coverage devoted to the deaths of Michael Jackson and Anna Nicole Smith.

The absurdity of it reminds me of *Cool Hand Luke*: "What we've got here is a failure to communicate."

The scope and scale of the climate changes are so vast, most people, especially those who aren't well informed, simply can't imagine what may be in store or what could be done about it. Could Colorado's pine forests really be wiped out in a matter of decades? Could Utah's Lake Powell really dry up? Could the West really lose a significant portion of its mountain snowpacks?

The Geography of Truth

So what can be done? The change in the journalism world should be seen as an opportunity. We need to create new organizations—likely, nonprofits—to help fund, promote, and practice environmental journalism and provide assistance to media outlets covering climate change. We need to find new ways to train young reporters about science and environmental issues and help them secure jobs in the new journalism world. We need to strengthen efforts to educate decision-makers in newsrooms about the emerging climate realities.

In the recent Gallup poll lies an optimistic note. Which group is most concerned about global warming and most likely to say the problem is underestimated? The young, those eighteen to twenty-nine. They have the most personally at stake in solving the climate crisis, so it's no wonder they are the ones most tuned in. As news consumers, this generation will not only want information about climate disruption, they'll likely demand it.

As our train chugs out of Glenwood Canyon, through historic Glenwood Springs, and into a western Colorado sunset, I'm envisioning a future conversation with Julia. "Dad, I remember the forests and all those animals. We didn't use to have all of these massive wildfires. What happened? Did anyone know what was going on?" Yes, they knew. "Why didn't anyone tell us?" I'll say, "What we had was a failure to communicate."

With the sun ducking down below the horizon, I ponder the implications of that hypothetical Julia/Daddy interrogation.

The West is a land of stark contrasts, with Utah leading the way. Hot and cold, light and dark, the confluence of the past and the future. One day it's a glorious bluebird sunshine day, the next comes a flash flood.

Survival in the West has often been on a knife edge, though today we seem to have utterly lost those lessons. Generations of Native Americans made a home in this often-inhospitable place. But sometimes environmental forces beyond human control won out, as the Ancestral Puebloans (also known as the Anasazi) learned painfully. Many experts believe severe drought may have been the last straw in their demise. It's a cautionary tale for those in dangerously thirsty, sprawling western cities like Los Angeles, Phoenix, and Las Vegas that are sucking water from the Colorado River.

Centuries after the Ancestral Puebloans, settlers came west, motivated by both survival and a sense of opportunity. Mormon pioneers arrived in what is now the Salt Lake Valley looking to establish their own Zion, the ideal nation, a heaven, utopia. Indeed, the West as a whole represents endless horizons, endless possibilities.

Along the Denver–to–Salt Lake route, we pass visual reminders of the many energy sources available to power the modern civilization that has been carved out of that world of possibilities. That energy fires the cars that pass on the interstate, the airplanes above, the homes and stores and factories, the laptops and the locomotives.

Out the Amtrak window, we see a glimmering solar array and natural gas wells, flares burning the sky, near Grand Junction. In southeast Utah, there are tailings from old uranium mines. Price, ten hours into the trip, is surrounded by coal mines, like Crandall Canyon, where nine miners died in 2007 pulling out the last pillars of black gold, the commodity that provides electricity to 50 percent of America and 95 percent of Utah. As we exit the last stretch of mostly open land and head down toward Provo and the heart of urban Utah, we pass something new: a wind farm with half a dozen stark white turbines, gently tilting like pinwheels.

With Julia now asleep on the top bunk, I'm thinking about Wallace Stegner, the writer and conservationist who famously described the essential optimism of the West as a "geography of hope." But to realize that hope, citizens in a democracy need to understand the reality of their own world. They need a "geography of truth," which only robust, thoughtful journalism can provide.

To accurately chart that geography requires the best possible information, from the scientists and the journalists, communicated to the public as clearly and compellingly as possible. Survival in the West and on this planet will depend on developing both the communication skills and lasting journalism institutions to match our technological prowess.

By the time the train arrives in Salt Lake, a couple of hours late, of course, there's a slight breeze blowing in the cool midnight air. It's a zephyr, a breeze from the west. That seems fitting. If the world is going to address the climate crisis, it's going to need some zephyr-like momentum, a gentle push that tempers both the fear and the hope with the truth.

John Daley is a broadcast journalist in Salt Lake City specializing in political, environmental, and investigative coverage. He also teaches journalism at the University of Utah and was a John S. Knight Fellow in journalism at Stanford, where he studied leadership in the age of global warming.

IT AIN'T YOUR FATHER'S FARMING— NEW MIND-SETS AND NEW PRACTICES IN THE AGE OF CLIMATE CHANGE
By Susan Moran

Barreling along in my Subaru Forester, I am on Interstate 70 heading east of Denver on this crisp, sunny day in February. The late-afternoon shadows extend their tendrils across field after field of closely groomed crops. Fellow Subarus—ubiquitous sightings in the Denver/Boulder metro area—become scarcer and scarcer, replaced by dusty white Toyota trucks and flatbeds carrying farming machinery. The flat topography is punctuated only by an occasional grain elevator and farmhouse.

I reach Exit 405 at the town of Seibert and steer south, passing the gas station/coffee shop at the intersection. I rumble along for five more miles, passing a few ranch-style homes and endless swaths of groomed fields on both sides, until I turn into a gravel driveway. Two black Labs bolt to the car, greeting me with yelps and licks.

I'm met at the door by Curtis Sayles, a graying man with a boyish face who's clad in jeans and a blue work shirt. He offers me a warm smile and a firm handshake. "I bet you could use some coffee," he says as he ushers me to a long wooden dining table in a spacious room and pours me a black brew. He sits down across the table and launches into talk about life on the farm and raising three daughters with his wife, Kerry.

I'm struck by the fact that Sayles, fifty-two, is as much an entrepreneur as an ordinary farmer. Or that "ordinary" farming looks a lot less ordinary as guys like Sayles scramble to keep the farms, and the towns they reside in, afloat by diversifying their income streams. Sayles, for example, is launching wind, biomass, and other ventures, and he is organizing farmers in the community to negotiate a lucrative contract with a large wind-farm developer that's been scoping out the area. If necessity is the mother of invention, then the people who work the land might be the ones to help us figure out how to save it. Just don't go calling Sayles a tree hugger.

Sayles is among a small but growing subset of farmers and ranchers who are becoming part of the climate-change solution. Many of them are driven more by a desire to save the farm than to save the planet. Some of them are generating renewable energy in the form of cooperative wind farms or biofuels production. Others, like Sayles, have entered carbon-trading markets, such as the voluntary Chicago Climate Exchange, as carbon-offset providers, as they're called in industry parlance. In essence, they are reducing or sequestering carbon dioxide in the soil through practices and technologies such as conservation tillage and methane digesters—a way to capture this superpotent greenhouse gas on dairy farms. Still other farmers are planting forests or native grasslands where they had grown crops. The climate exchange, as well as other regional carbon-offset programs, are gaining appeal among farmers from Georgia to Oregon.

Warmed by coffee, we soon jump into Sayles's 1980 Ford F-150 truck and head out to survey his 5,000 acres. The wind gusts and whistles loudly through the door cracks, making it tough to carry on a conversation. Dust clouds trail us down the narrow dirt road. Sayles pulls up to stop at the border of his and his neighbor's fields. The contrast between his fields and those surrounding them is stark: one side is tidy, groomed, slick, while the other is chaotic, spiky, messy.

Farmer Curtis Sayles walks among the weed-covered and unused traditional tilling equipment on his father's farm in eastern Colorado. Courtesy of Kevin Maloney

We step down from the truck and walk a few yards onto his field. There's no dirt blowing across the ground. Sayles presses his brown, round-toed cowboy boot into the spongy soil for signs that his winter wheat is starting to peek out. New green shoots are difficult to see because the field has been covered with a helter-skelter splash of decaying sunflower stalks and corn husks. It looks like a bad hair day on a grand scale. It's whimsical and wild. For Sayles, it's just practical. He never plows his fields, on which he plants a continuous rotation of winter wheat, corn, sunflower, and millet. His method is known as no-till. In the eyes of some of his tractor-driving neighbors, it approaches heresy.

"We don't go to the coffee shop much because I don't like people snickering at me," he says, nodding toward his neighbor's field as he dips both hands into his pockets. "I think we're doing things that don't fit into their mind-set."

He was referring to the restaurant I passed just off Interstate 70 up the road, Seibert Travel Plaza. It's the only coffee shop left in this sleepy town. In fact, it's the only social hub besides High Plains High School, where Sayles keeps score and Kerry coaches for the girls' basketball team, and where most of the town's 150 residents come to cheer and trade gossip.

Despite neighbors' barbs, Sayles may have the last laugh. Traditional plowing, or tilling, helps get rid of weeds and generally makes using

Curtis Sayles and other farmers are using more environmentally sensitive techniques to raise cash crops and receive carbon credits. Courtesy of Kevin Maloney

fertilizers and pesticides easier. But unlike plowing, the no-till method shields the soil's carbon-rich humus from contact with oxygen in the atmosphere. Consequently, the humus doesn't decay as quickly.

Sayles sees how the wind blows away his neighbors' topsoil, perhaps a farmer's most precious asset. By contrast, his topsoil stays put, protected by the residue from previous harvests. The blanketing also keeps his soil moist and rich with yield-boosting organic matter called humus. And it helps prevent water erosion, which is critical in this arid region where farmers like Sayles rely on dryland farming. What's more, no-tillage lets Sayles spend less time behind the wheel of his tractor plowing and less money filling the tank with diesel fuel.

Lately he has begun to reap another reward. One day last summer, he pulled out of his mailbox a $6,200 check from the National Farmers Union. It came on behalf of the Chicago Climate Exchange, a capitalist's approach to stemming climate change. The checks, issued to farmers and ranchers enrolled in the program, actually come from faraway utilities, manufacturers, and other corporations that are attempting to assuage their guilt as polluters by paying carbon-offset providers like Sayles to sequester carbon in soil and plants. Wealthy companies can pay their way to faster reductions in greenhouse gas emissions when they can't or don't want to make dramatic cuts on their own.

Curtis Sayles talks with his brother Scott among the mixed stubble of sunflowers, corn, and wheat on his land in eastern Colorado. Courtesy of Kevin Maloney

Sayles, who started farming in Seibert with his father in 1980 after a stint in Houston working in the oil industry, has never taken farming for granted. He always has looked for new ways to boost his yields, nourish the soil, and generate more income. So he figured, why not enroll in the Chicago exchange's carbon-credit program? "I wouldn't say I'm an environmentalist," he says. "To be candid, we're getting paid to do something we're already doing. It was a no-brainer."

To Sayles, the $6,200 could have gone toward financing two of his daughters' weddings this year, or toward a $30,832 John Deere no-till drill, or toward his dwindling retirement fund. To a large power company like American Electric Power, which made $14.6 billion in revenues in 2008, that same $6,200 might help furnish an office with a few Herman Miller chairs at $620 each.

To the voluntary Chicago Climate Exchange and regional carbon-trading schemes that have cropped up in the country, no-till farming is one of several agricultural and land-management practices that qualify as carbon offsets. That's because soil loses most of its carbon content during plowing, which in turn releases carbon dioxide gas into the atmosphere.

Carbon offsets, or credits, produced through agricultural practices—particularly no-tillage—have been met with skepticism by some economists, scientists, politicians, and others. They argue, for instance, that some practices don't amount to additional or new net reductions in greenhouse gas emissions, and that five-year contracts don't guarantee permanent emissions reductions and should thus not be considered an emissions offset. If the carbon is eventually released back into the atmosphere, does sequestering it amount to any more than a feel-good measure?

Pieter Tans is a senior scientist at the National Oceanic and Atmospheric Administration's (NOAA) Earth System Research Laboratory in Boulder, Colorado. The lab runs several carbon dioxide measurement systems around the world, including the Mauna Loa observatory in Hawaii. Tans says there is no evidence that carbon-offset programs or markets, including the European Union's Emission Trading Scheme, have put a squiggle in the atmospheric carbon dioxide data, largely because these mechanisms are still so new and small. Further, he adds, any blip in the data would be wiped out by China's building of at least one coal-fired power plant each week.

"Looking from the atmosphere, we cannot say at the moment that they're making a dent because the amounts we're talking about are so small," says Tans.

But on a smaller scale, recent NOAA studies of Iowa cropland have shown that a land-based instrument protruding into the air detected a surprisingly large amount of carbon dioxide uptake by corn and soybean plants from the atmosphere.

Tans estimates that if carbon could be reliably stored in soils for at least fifty years and then released into the atmosphere, that time delay could make a positive difference in buying time for effective climate-change policy and more advanced measuring technology to emerge. Meanwhile, he says, farming and ranching practices that sequester carbon are worthwhile because they yield positive environmental and economic benefits.

Farmers like Sayles are getting paid for continuing not to plow their croplands. Some people view it as an unworthy subsidy—paying someone to do nothing different. And it disrupts the time-tested notion that doing nothing (namely, letting nature take its course rather than taming, or grooming, it into submission) can't be considered labor, and thus should not be valued. Nature follows its own logic; sometimes the best thing to do is leave it alone. However the scientific and accounting questions get resolved through carbon markets, it's worth asking ourselves, who should decide who gets rewarded and who should pay, and how do we deal with the trade-offs? These questions are coming to the forefront as Congress, under the Barack Obama administration, begins shaping long-anticipated climate legislation. The most popular proposal has been a cap-and-trade system.

In any case, agriculture and other land practices will likely play a role in such a system. Globally, agriculture and ranchlands contribute at least 12 percent of greenhouse gas emissions, and roughly 8 percent of US emissions, making them a significant climate-change culprit. That amounts to less than the planet-warming emissions from the electricity and transportation sectors, for instance, but many scientists, policy makers, and economists view agriculture as a low-hanging fruit. They say that reducing a significant amount of carbon dioxide and other offending gases by sequestering carbon in soils and vegetation can be done faster and cheaper than, say, injecting carbon dioxide from coal-fired power plants into deep rock formations.

"The bottom line is there's plenty of scope for agriculture to participate in carbon offsets," says Keith Paustian, a soil ecologist at Colorado State

University at Fort Collins. "And wise practices bring not only greenhouse gas benefits, but other benefits, like less nutrient pollution and less soil erosion."

One environmental trade-off of practicing no-tillage is that with some crops it requires more spraying of Roundup and other chemical weed killers in lieu of plowing before planting. But the environmental benefits outweigh the downside, says Karen Scanlon, executive director of the Conservation Technology Information Center in Lafayette, Indiana. The added crop detritus that's left on the soil, especially with continuous no-tillage, keeps the soil in place so the chemicals don't blow away with the soil. (Advances in technology, such as global positioning system auto-steer sprayers, have also allowed conventional and conservation-tillage farmers alike to apply herbicides and fertilizers much more precisely and efficiently than they did years ago.) The residue from no-tillage also improves water quality of the soil and provides food for wildlife, making the surrounding habitat richer, Scanlon said.

Back on his farm in Seibert, Sayles wonders how the dry, windy winter in eastern Colorado will play out. "Nobody really knows what's going on with the climate—whether it's global warming or what," he says. "But I feel even better we're doing continuous no-till because there's always residue on the ground, no matter how strong the wind blows. I'd say we've got our bases covered."

Susan Moran is a freelance journalist who writes for *The New York Times*, *The Economist*, and other publications. She is currently on a Knight Science Journalism Fellowship at the Massachusetts Institute of Technology.

DEAD TREES
By Jim Robbins

On the side of a mountain on the outskirts of Montana's capital city, loggers are racing against a beetle grub the size of a grain of rice.

A grinding machine called a feller-buncher grabs a pine tree by the trunk and with a large jaw snips it off at the bottom, while another piece of equipment grabs the tree in a metal claw and in a spray of wood splinters strips off the limbs. Within a few minutes, a living tree has become a log, ready for a sawmill.

While logging has long been part of the Rocky Mountains, it is taking on new urgency. From New Mexico to British Columbia, the region's signature pine forests are succumbing to a massive infestation of beetles inexorably munching their way through the mountains, turning a blanket of green forest to a blanket of rust red.

A logger uses a feller-buncher to gather beetle-infested trees after they were cut down on private property outside Helena, Montana. (Courtesy of Anne Sherwood)

"It's decimating trees," says Gary Ellingson, a forestry consultant for Northwest Management who is advising on the emergency sale of private land here. "I've literally had people in my office crying. People are powerless."

Foresters say the historic outbreak is caused by a perfect storm of events. First, because fires have been suppressed for so long, all forests are roughly the same age and big enough to be susceptible to beetles. Second, a decade of drought has weakened the trees. Third, typical hard winters have softened, which allows the beetles to flourish and expand their range.

In hopes of keeping their forests from completely dying, to earn money by selling dead and infected trees, and to mitigate fire risks, landowners are scrambling to cut down huge swaths of forest. If enough are cut—up to 75 percent—those left behind might survive.

In northern Colorado and southern Wyoming, the unprecedented acreage of standing dead forest has become a crisis, and the US Forest Service has set up a special incident-management team to protect homes and communities across the West from dead forests that have exacerbated the threat of fire.

The dying of the forests now has more impact because the last twenty years have also seen large numbers of people building homes in the forest.

One of the hallmarks of the outbreak is the vexing lack of methods to stop the beetle. "The Latin name is *Dendroctonus*, which means 'tree killer,'"

A mountain pine beetle, about the size of a grain of rice, clings to the end of a pocket knife after being removed from a pine tree just south of Bozeman, Montana. (Courtesy of Anne Sherwood)

says Gregg DeNitto, a forest service entomologist in Missoula, Montana. "They are very effective."

The black, hard-shelled insect, the size of a fingertip, drills through pine bark and digs a gallery in the wood, where it lays its eggs. When the larvae hatch under the bark, they eat the sweet, rich cambium layer that provides nutrients to the tree. They also inject a fungus to stop the tree from moving sap, which could drown the larvae. That fungus stains the wood blue.

To fend off the bugs, trees emit white resin, which looks like candle wax, into the beetle's drill hole. Sometimes the tree wins and entombs the beetle. Often, though, the attacker puts out a pheromone-based call for reinforcements and more of them swarm the tree. In a drought, the tree has trouble producing enough resin and is overwhelmed.

There are some defenses. Tree owners nail something called aggregator pheromone in a small packet to a tree, which mimics the chemical scent given off by beetles when a tree is full of insects. It can work when beetles aren't too numerous, but at some point the beetles aren't deterred.

So-called high-value trees, big old trees that shade campgrounds or yards, can be sprayed with an insecticide. But the trees need to be sprayed from the base to where they are less than four inches around, and it's expensive. Each tree costs about $10 to $15 if hundreds are sprayed.

As things get bad, enough logging is the only option, selling trees before they die or rot. The economics don't pencil out well, though, largely because dead trees have a lower value, the decline in home building has driven down the price of lumber, and the cost of fuel is high.

The beetles, experts say, will have their way and only be checked if bitterly cold weather returns to the Rockies—which hasn't happened since the 1980s.

While Montana has seen a million acres killed or dying, in northern Colorado and southern Wyoming the crisis in the states' lodgepole pine forests is historically unprecedented.

"We're seeing exponential growth of the infestation," says Clint Kyhl, director of an incident-management team in Laramie, Wyoming, set up to deal with the crisis caused by the huge swath of dead forests in northern Colorado and southern Wyoming. In 2006, there were a million acres of dead trees. In 2008, it was 1.5 million. In 2009, it is expected to total over 2 million.

In the next three to five years, Kyhl says, virtually all of Colorado's lodgepole pine trees over five inches in diameter will be lost, about 5 million acres. "Already, in many places every lodgepole over five inches is dead as far as the eye can see," he says. (There's not enough food in a tree to sustain the beetles if it's less than five inches tall.)

Lodgepole pines are largely confined to high altitudes. But the beetles have moved into ponderosa pine forests on Colorado's Front Range, Kyhl says, which means they could kill forests around homes in the densely populated region.

In the Canadian provinces of British Columbia and Alberta, the problem is most severe of all, the largest known insect infestation in the history of North America, according to officials. British Columbia has lost more than 34 million acres of lodgepole pine forest, and a freak wind event last year blew beetles over the Continental Divide to Alberta.

Cold weather always kept the beetles from crossing over. To keep the bugs in check, several days of temperatures that touch forty below zero are needed. But warmer temperatures and the wind changed the beetles' range, and experts fear they could travel all the way to the Great Lakes.

So many trees have died in British Columbia, and so much carbon from the dying trees has been released into the environment, that experts say the forests have gone from a carbon sink to a carbon source.

The death of the forests worries the tourism industry across the West. Because of the hazard of falling trees, many ski areas have had to cut down their forests and revegetate. At Vail Resort, for example, which has been particularly hard-hit, workers have removed thousands of dead trees and planted new ones.

The dead trees that blanket the mountains are shifting ecosystems as well. In Yellowstone, for example, the beetles are killing the whitebark pine trees, which grow nuts rich in fat that are critical to grizzly bears. Biologists say streams will flood because live trees will no longer catch snow and allow it to slowly melt, and thus the epidemic could injure salmon.

The biggest concern from the millions of acres of dead forests across the West, though, is wildfire. The forest service incident management team is charged with trying to protect homes, communities, and critical infrastructures from fires in dead forests. "It's like burning your Christmas tree," says Kyhl. "The chance for catastrophic fire is increased."

The scale of the problem is so big, Kyhl says, that the focus has been on protecting communities surrounded by dead forest.

"The fire risk has increased exponentially and we feel like a box of matches waiting for someone to light one and ignite the box," says Ray

Jennings, director of emergency management for Grand County, where the number of dead trees is staggering.

After the trees die, the risk of crown fires, which move through the canopy, is high. But when the needles fall off standing trees after a few years, the fire risk is less severe. After four or five years, as the trees fall to the ground, like a huge pile of jackstraws, the threat of catastrophic fire is most worrisome.

Dead trees create a lot of fuel on the ground, Kyhl says, so fires could be unusually intense and severely damage soils, preventing the next generation of forest and causing massive flooding.

Rain on damaged ground could lead also to widespread mudslides and silt buildup in rivers and reservoirs, which many mountain communities depend on for water. Strontia Springs Reservoir, a main water source for Denver, required a $20 million cleanup after a large fire resulted in severe erosion.

Some towns, such as Steamboat Springs and Vail, Colorado, are surrounded by dead forests so the forest service and logging companies are clear-cutting "defensible space" so firefighters have a place from which to fight fires.

Jennings has been working on emergency plans to evacuate towns and resorts in the event of a fire in dead and downed forests, which can move faster than fires in a green forest. Part of the national power grid goes through Grand County and the aluminum towers could melt in the intense fires, so officials will cut trees near the towers.

Mountain pine beetles burrow into the bark and girdle the tree. If the tree can express enough sap to push out the beetles, it can survive, but years of drought have weakened stands and left them vulnerable. The beetles leave a blue stain on the wood, making it less valuable as lumber. (Courtesy of Anne Sherwood)

The other major problem is large numbers of falling trees. Across the West, officials have closed campgrounds for fear trees could fall on campers. They must be logged before they can be reopened. All but fourteen have been reopened.

But there's a lot more to do. "We know they are going to fall," says Kyhl. "And they are going to fall in the next ten to fifteen years. There are campgrounds, thousands of miles of roads, picnic areas, power lines, and trails. How do we keep the facilities open for people to use?"

The agency is faced with the task of clearing a strip of 75 to 100 feet of dead trees along highways so they aren't closed by blowdowns.

Then there is a question of what do with the wood. Sawmills have diminished in the West in recent years so there aren't enough mills to take all of the timber.

In Colorado, entrepreneurs have been scrambling to find ways to use it. Two new pellet plants have been built, which turn the trees into sawdust and then pack them into a clean-burning pellet used in woodstoves.

Some trees are being shredded for use in biomass boilers, and carpenters are using the pine stained blue from a fungus left behind by the beetle for such things as furniture.

In Alberta, a newsprint mill is testing a new system to use the millions of dead acres of pines for paper. Because of the fungal stain, the trees aren't bright enough to be used for paper without being treated, so a computerized process adjusts the amount of bleach that is added.

Still, the volume of timber used is small compared to the vast acreage of dead trees.

Meanwhile, the West that depends on tourism wonders what their customers think about the dramatic change in scenery. Four million visitors a year come to Grand County in Colorado to recreate and sightsee. "What happens," Jennings asked rhetorically, "if this becomes an ugly place to be?"

Jim Robbins is a freelance journalist in Helena, Montana, where he has written for *The New York Times* for more than twenty-five years. He is also a frequent contributor to *Condé Nast Traveler*, for which he has written environmentally themed travel stories on Peru, Chile, Mongolia, Sweden, Mexico, and numerous other places. He has written three books, most recently on the critical role of the process of attention in human physiology and psychology.

GRAND COUNTY—
LIFE AMONG THE PINE BEETLES
By Hillary Rosner

Back in 1997, when Charles and Nancy Henry moved full-time to their 38.5-acre property just outside the town of Granby, Colorado, the views were just about perfect. At 8,600 feet, with aspen groves behind the house and 360-degree lodgepole pine forest vistas, it was Charles Henry's version of paradise. A semiretired agricultural and natural resource reporter, Henry set about restoring his patch of long-neglected, largely overgrown forest to good health—thinning lodgepole pines here, doing small clear-cuts there, removing mistletoe (an invasive weed), and helping create a more diverse ecosystem to replace the often homogenous stands of lodgepole. "It was a gorgeous place to live," Henry said. "I can tolerate the winters, and the summers are wonderful, and every way you looked was something pretty. It was something we worked very hard for and finally realized."

Today, though, the Henrys are stewards of acres of wasted trees, and their mountain retreat is a lookout on a natural—or seminatural—disaster. Up and down the Rocky Mountains, an unstoppable plague of beetles, spurred on by rising temperatures and homogenous forests, is ambushing pine trees. "Every day I look out and say, 'Ugh.' I just hate it," said Henry. "I thought I'd get over it, but I haven't." As mountain pine beetles rampaged through Grand County over the past five years, laying siege to nearly 550,000 acres of lodgepole pines that were overcrowded from decades of fire suppression, Henry's focus on his own property turned from forest stewardship to wildfire mitigation—trying to lower the risk of a catastrophic wildfire—and ensuring that the next-generation forest is a healthy one, even if he won't be there to see it. His remaining twenty-one acres of pine forest will soon be largely cut, to start the regeneration process. It's this latter phenomenon that still brings Henry small moments of joy when he walks and works his land.

"Look at all the little lodgepoles!" he exclaimed on a cool May morning while standing on his muddy road to gaze at a small area that he clear-cut seven years ago to control mistletoe and combat a porcupine problem ("Technically, you could shoot 'em, but I believe in live and let live"). A green island of swaying, breathing trees in a brown ocean of rigid, spiritless ones, Henry's patch of clear-cut teemed with eight-foot aspens and a flourishing crop of knee-high baby lodgepoles. The miniature trees look so bright and hopeful, it's hard not to smile when you see them. Which is what Henry did, for a few moments at least, until he looked up the slope again. "Our goal is to get these dead trees out of here before one tree falls and knocks out three little healthy ones."

These days in Grand County, a haven of mountains, rivers, lakes, and lodgepole-dominated forests in north-central Colorado that includes a portion of Rocky Mountain National Park, the impacts of the mountain pine-beetle epidemic are as much a fact of daily life as snow in winter. The first part of Colorado hit by the beetles, it's also the farthest along on the trajectory of forest death, and residents are coping with everything from lack of privacy as their now-treeless homes are suddenly exposed, to the financial burden of tree removal or spraying. There's a low-grade paranoia about trees falling— on homes, power lines, boats, even people. In October 2008, a beetle-kill pine fell and killed a logger from Granby as he was removing slash in the town of Grand Lake. Earlier the same month, another man was knocked to the ground but survived when a dead lodgepole toppled onto a county road.

"It's like dealing with a disaster that moves slowly and never goes away," said Craig Magwire, district ranger for the US Forest Service's Sulphur Ranger District. "As opposed to a fire event, where you bring in resources and work the fire and then you go home, here, it's continuous."

Magwire spends much of his day troubleshooting beetle-related problems on the public's land, and then he goes home to a wooded subdivision just north of Granby, where he and his neighbors have each been assessed thousands of dollars for their homeowner's association beetle-mitigation efforts—selective logging and aggressive pesticide spraying. The effectiveness of spraying is a subject of much debate in these parts; to work, the chemicals must cover 100 percent of the tree, so a shoddy job or a too-tall tree can mean wasted money and needlessly dispersed poison, which some believe is killing songbirds. One plus side of the epidemic, though, has been a noticeable spike in the local woodpecker population, which feeds on the beetles.

On a wall in the forest service's conference room, a large map dotted with dark areas shows 16,000 acres where trees were cut in an effort to create firebreaks. Some of these are deep in the forest; others are where the forest meets the towns. The overwhelming concern in Grand County is the threat that an uncontrollable wildfire will rage through the desiccated forests, engulfing homes and businesses—and possibly people—and leaving a trail of tragedy. With the firebreaks, the forest service is trying to up the chances that a wildfire won't prove devastating—but, as Magwire himself admitted, they're just tweaking the odds. The Sulphur District contains 185,000 acres of lodgepole-dominated forests, most of which are already dead and could soon fall, producing a dangerous mass of fuel on the forest floor. "We couldn't remove trees on all those acres even if we wanted to," Magwire said. "The money isn't there to do that, and there isn't an industry there that could use the material."

As the region goes into what Winter Park's town manager, Drew Nelson, calls "full-on clear-cut mode," what to do with all that wood is becoming an ever-tougher problem. The recession has only made things worse by halting home building and stalling sales of timber products. The forest service holds timber sales on 1,000-acre parcels; every Sulphur District sale for the past eight years has found a buyer. But the downturn in the housing market could change that—which in turn would curtail Magwire's forest-thinning operations and potentially tip the catastrophic-wildfire odds back in the fire's favor. "The biggest concern today is, if we don't have a timber industry, how are we going to move material out of the woods?" said Magwire. "In order for us to do our job, we need to have a market."

But while the forest service has so far been able to profit from its timber, homeowners, subdivisions, and even town governments are spending a fortune to cut down their trees. Small businesses have opened up to make furniture and paneling out of the beetle-kill wood, which is stained a bluish shade from a fungus the beetles inject into it. (The window for using the trees for structural products, such as two-by-fours, is brief once the beetles invade.) But these companies, too, have been hobbled by the economy, and even in boom times they could only utilize a fraction of the timber coming down. A new organization, the Colorado Beetle Kill Trade Association, is hoping to speed the creation of profitable new uses for the wood. For homeowners, the cost to have a tree cut and hauled away varies from $100 for just one or two trees—particularly right next to a house, where the logger might have to take it down in sections—to $20 apiece for 1,000 trees. Henry, who sought estimates to cut, limb, stack, slash, and remove timber from his land and that of his neighbors, received bids as low as $800 to $1,200 an acre, which works out to roughly $2 per tree. "The market's flooded," he said. "So that part has swung in our favor a bit. I wouldn't be considering it if it was still $20 a tree. I mean, that would cost me $250,000!"

The town of Winter Park uses funds from a mill levy to collect trees that residents remove from their property. A crew of foresters hauls the wood five days a week to an air curtain burner, which combusts it at high speed and without spewing smoke into the valley. "It'd be great if we could've figured out some way to harness all of this energy that's just escaping," said Nelson, "but unfortunately the epidemic was just so fast that there really wasn't this great plan—'Oh, wow, we could put a wood-burning stove in every house and it would heat the county for years to come.'"

According to Russ Chameroy, Winter Park's public works director, only two companies within two hours of the town are still taking in wood; both are mills that produce wood pellets for pellet-burning stoves, and both have pellets stacked as far as you can see. One mill reportedly has enough pellets for 175 years' worth of burning.

On a perfect blue-sky Colorado morning, Bruce Van Bockern stopped his truck on a hill overlooking Grand Lake. The panoramas were spectacular—and new. Houses that once faced woods now have unobstructed lake views. "You couldn't see a house up here two years ago," said Van Bockern, operations manager for Mountain Parks Electric, the local power co-op. He drives through this area frequently to check on crews cutting trees along the company's power lines. Some days he suddenly doesn't quite know where he is. "I've gone into some subdivisions that you don't even recognize because there are no trees," he said. One local logging company has even begun using the slogan "Creating new views."

Van Bockern, whose sons capitalized on the beetle infestation with summer logging jobs, spends the bulk of his time trying to ensure that trees don't fall on power lines. Of the co-op's 1,388 miles of overhead power lines, 460 miles are bordered by trees. But it's not the dead trees Van Bockern worries about. Lodgepoles live in groups, where their numbers help protect them from the wind. But when only a few live ones remain, there's nothing to shield them. Compounding the problem, the dead trees don't soak up water—or hold snow—so spring runoff causes excess soil moisture, weakening the live trees' roots. Van Bockern stopped to examine a beautiful two-foot-wide lodgepole that was leaning precipitously toward a nearby house. On the ground, a break was already visible in the soil where the tree's roots were coming loose.

Up the road, in a subdivision called Woodpecker Hill, the sound of chain saws filled the air. Groups of dead lodgepoles and spindly-looking live ones still towered above some houses, but everywhere the ground was littered with fallen timber. An early May windstorm that blew through here sent scores of trees crashing to the ground; dead ones snapped, live ones were uprooted, several toppled on cabins and outbuildings. Paul Shelley, an electrician who lives on a corner lot, was out doing never-ending yard work: burning a slash pile and plotting an outline for a log border around his property. "We use the wood any way we can," he said. A few feet from the side of Shelley's house, a lone lodgepole was still bushy and green—except for the top, where death was creeping in. A band of blue ribbon marked it for pesticide application and showed that even though many people here have finally resigned themselves to clear-cutting, emotional attachments to beloved trees linger and prevail. "We've sprayed this tree for years," Shelley said. "This is the one tree my wife's real sentimental about."

It's a feeling Magwire understands. "You get to that point where you have to decide," he said. "Do you want to continue to spray or just take out all the trees? It's a science decision and an emotional one and a financial one. It's all those things."

Not all the beetles have wrought is bad. Allergy sufferers are breathing easier thanks to less pine pollen in the air. Sales of chain saws and related merchandise—chains, chaps, hard hats—are up. Free firewood is everywhere. People have learned to revere healthy forests in a way they didn't before—and are figuring out how to better manage them.

But then there are the daily nuisances: fallen trees blocking roads, basements flooding from increased soil moisture, finger-pointing among neighbors over whose property lines the beetles crossed. Concern is rising over impacts on the recreation industry that's an integral part of the county's economy. "If you have a choice between a green forest and a dead forest, where would you build your million-dollar resort?" mused Ron Cousineau, district forester for the Colorado State Forest Service's Granby district.

"Things are going to become a lot more difficult to do out in the woods," said Nelson, whose town economy could be hit if hikers, bikers, and hunters choose greener pastures for their outings. Property owners like Charles Henry have been wondering about this too. "Are people going to want to come hike when all the trails are blocked or there's danger of getting killed by a falling tree?" Henry asked.

The crisis isn't limited to Grand County; trees in Summit and Routt counties are also being decimated. But even Grand's problems resonate beyond the immediate area. Denver Water, the water utility for 1.3 million residents of Denver and the surrounding suburbs, gets a quarter of its supplies from a collection system that includes the county. Wildfires lead to erosion; erosion clogs reservoirs with sediment that costs millions of dollars to remove. "It causes the soil chemistry to change," said Don Kennedy, an environmental scientist with the utility's planning division, "so you get these hydrophobic soils that repel water." When heavy rains come, the denuded landscape, coupled with the water-repellent soil, causes water flows that bring huge amounts of erosion. Denver Water is working with a long list of organizations—including Winter Park and the forest service—to prioritize key areas for logging. "It's a lot cheaper to address it now than later," said Kennedy. "Nobody wants to spend $25 million on removing sediment out of the reservoir."

If there is a silver lining, it's that the pace of Grand County's outbreak seems to be slowing compared to other areas of the state that are a few years behind. Once the mature trees are all dead, the beetles will have nothing left to feed on—and hopefully their populations will crash. Nelson believes the area faces four more years of cutting, hauling, and fire mitigation—and that a huge wildfire somewhere in central Colorado is all but inevitable. But several decades from now, Grand County's slopes might once again be covered in living forests, with diverse stands of pine, aspen, spruce, and fir that can better withstand another onslaught.

Henry likes to envision that day. "We're not going to live here forever," he said of his former paradise. "My goal would be to put it in a land conservancy and sell it to someone who'll keep it as forest. I'm hoping that by spending the money to remove the dead trees and get this forest process started sooner, in five to ten years when I'm ready to sell it, I'll have a more desirable property than someone who hasn't done it. I'm trying to be altruistic about the forest value, but there's a mercenary, a dollar and cents factor underneath."

Hillary Rosner has written for *The New York Times*, *Mother Jones*, *Men's Journal*, *Popular Science*, *Seed*, *Audubon*, *High Country News*, *Slate*, *Grist*, and many other publications, and she is the coauthor of the book *Go Green, Live Rich*. She also contributed to Al Gore's book *An Inconvenient Truth*. She holds a master of science in environmental studies from the University of Colorado at Boulder, where she studied on a National Science Foundation fellowship.

WHAT'S KILLING THE ASPEN?
THE SIGNATURE TREE OF THE ROCKIES IS IN TROUBLE
By Michelle Nijhuis

It's a relentlessly sunny day in the Rocky Mountains, and here at 9,000 feet, on the Grand Mesa in western Colorado, the aspen trees should be casting a shadow. But something is wrong in this stand: the treetops are nearly bare, their branches twisting starkly into the blue sky. Sarah Tharp, a wiry biologist for the US Forest Service, hoists a small ax, takes aim, and delivers an angled blow to an aspen trunk, peeling off a sample of diseased bark. "Sometimes," she says, "I feel like a coroner."

Aspen, one of the few broad-leaved trees to grow at high altitude in western mountains, are emblems of the Rockies. Their lean, chalky trunks are instantly recognizable on an alpine slope, their blazing-yellow fall displays part of the region's seasonal clockwork. The characteristic flutter of their heart-shaped leaves in the breeze gives them their nickname—"quakies"—and fills their stands with an unmistakable *shhhhh*.

In 2004, foresters noticed that aspen in western Colorado were falling silent. While the trees have always been susceptible to disease and insect attacks, especially in old age, "this was totally different from anything we'd seen before," says forester Wayne Shepperd. "In the past, you'd maybe see rapid die-off of one stand out of an entire landscape—it wasn't really a big deal. But now, we're seeing whole portions of the landscape go."

By 2006, close to 150,000 acres of Colorado aspen were dead or damaged, according to aerial surveys. By the following year, the grim phenomenon had a name—sudden aspen decline, or SAD—and by 2008, the damaged areas had exceeded half a million acres, with 17 percent of the state's aspen showing declines. In many places, patches of bare and dying treetops are as noticeable as missing teeth, and some sickly areas stretch for miles. Aspen declines are also under way in Wyoming, Utah, and elsewhere in the Rockies. Surveys of two national forests in Arizona showed that from 2000 to 2007, lower-elevation areas lost 90 percent of their aspen.

Aspen grow in clones, or groups of genetically identical trunks. Some clones are thousands of years old, although individual trees live 150 years at most. One especially large stand in Utah, known as Pando, after the Latin for "I spread," was recently confirmed by geneticists to cover 108 acres. It is variously said to be the world's heaviest, largest, or oldest organism. Disturbances such as wildfires or disease usually prompt clones to send up a slew of fresh sprouts, but new growth is rare in SAD-affected stands.

Tharp and three other young forest service biologists—under the genial supervision of veteran plant pathologist Jim Worrall—are chasing down the causes of the decline. They walk among the aspen trunks and divvy up their tasks for the day.

"You want me to dig? Is that where this is heading?" Worrall teases the crew members, who are outfitted in hard hats and orange vests and sport the occasional nose piercing.

A tiny mark on the bark of one trunk prompts Angel Watkins to probe underneath with a knife, where she finds that the wood is decorated by the convoluted tracks of bronze poplar borer larvae. While the inch-long larvae don't usually kill aspen outright, their trails can weaken the trees and open new portals to fungal infections, which in turn form oozing bruises under the bark. On another tree, Worrall finds small cracks like those on the surface of a cookie, a clue that tunneling underneath has dried out the bark. Closer inspection turns up a bark beetle, no more than one-twelfth-inch long but capable, en masse, of cutting off the tree's nutrient supply.

"These beetles are the biggest mystery," says Worrall. Before SAD, aspen bark beetles were known to science, but "most entomologists who worked on aspen had never heard of them," he says. His crew now finds bark beetles in almost every damaged stand. They've also observed that some fungi, borers, and other insects and diseases are proliferating.

The most extensive SAD is in the hottest and driest areas—low-lying, south-facing slopes. The pattern suggests that the region's extreme drought and high temperatures—both possible symptoms of global warming—have weakened the trees, allowing more disease and insect attacks.

It seems that new stems aren't growing back after trees die because drought and heat have stressed the trees. During drought, aspen close off microscopic openings in their leaves, a survival measure that slows water loss but also slows the uptake of carbon dioxide, required for photosynthesis. As a result, the trees can't convert as much sunlight into sugar. Worrall speculates that the trees absorb stored energy from their own roots, eventually killing the roots and preventing the rise of new aspen sprouts. "They basically starve to death," he says.

The drought here has lasted nearly a decade, and climate scientists predict that severe droughts will strike even more often in parts of the West as greenhouse gas levels continue to rise and contribute to global warming. "If we have more hot, dry periods as predicted, SAD will continue," says Worrall. Aspen at lower elevations will likely disappear, he says, and those at higher elevations will be weaker and sparser.

Aspen aren't the only trees in trouble in the Rockies. The needles of many spruce and pine trees in Colorado are tinged with red, a sign of bark beetle infestation. The outbreak began in 1996, and today 1.5 million acres are infected. Foresters recently projected that the state will lose most of its mature lodgepole pines to beetles within the next five years. Whitebark pines, whose fatty seeds provide meals for grizzly bears in the northern Rockies, have long been protected from insect attack because they thrive in high-mountain habitat, but invading beetles have now knocked out most of the mature trees. Biologists say several types of bark beetles are reproducing more quickly and expanding their range, thanks to warming trends that allow the insects to survive winters at higher elevations and more northern latitudes.

"We're seeing major ecological responses to warming," says Thomas Veblen, an ecologist at the University of Colorado at Boulder and a longtime student of Rocky Mountain forests. "That's the common theme that's hitting everybody in the face."

As Worrall and his crew of biologists continue to investigate the damage done by SAD, a pair of Stanford University graduate students, Bill Anderegg and Alexander Nees, are using satellite data and historical climate records to take a closer look at the causes of aspen decline. Dan Kashian, a biologist at Wayne State University in Michigan, is studying the effect of drought on aspen in both Colorado and Minnesota.

Meanwhile, the forest service is testing treatments for the decline. In some places, researchers find, logging and controlled burns can be used to encourage aspen stands to generate new trees. In northern Arizona, foresters have fenced off several hundred acres of aspen in the Coconino National Forest, hoping the barriers will protect new growth from hungry elk and deer. But no one has found a cure.

In the fall, aspen's golden foliage creates a stunning contrast with the surrounding evergreens. These dramatic panoramas appear to be threatened. Future visitors to the Rockies are likely to find an altered forest, if, as experts foresee, aspen cede territory to evergreens or open meadows. Not that a forest is ever a static thing. "The forest of our grandparents' time wasn't the best of all possible forests, ours isn't the best of all possible forests, and the forest of the future won't be, either," says Dan Binkley of the Colorado Forest Restoration Institute at Colorado State University. Still, aspen's grandeur would be sorely missed.

Michelle Nijhuis is a contributing editor of *High Country News*. Her work has also appeared in *Smithsonian*, *National Geographic*, *The New York Times*, and the anthologies *Best American Science Writing* and *Best American Science and Nature Writing*. She and her family live off the grid in Paonia, Colorado.

THE NEW IMPERATIVE—
LAND CONSERVATION AND CLIMATE CHANGE
By Tim Sullivan

The environmental, conservation, and natural resource professionals of the twenty-first century confront the two greatest challenges these professions have ever faced: first, to help reduce the human-caused emissions of greenhouse gases responsible for causing climate change, and second, to prepare natural and human systems to withstand the inevitable changes in climate we will face in the future. Even with aggressive and unprecedented efforts to change the ways we produce and use energy, the ways we move ourselves and goods, and even the ways we build our cities and feed ourselves, natural environments are likely to face changes in climate conditions of a magnitude and pace undocumented in the historical record. What makes the challenge of preparing for climate change even more difficult is that we start in arrears, in a conservation deficit created by the changes that humans have wrought on species and ecosystems in the past.

The need to prepare now for future climate change, combined with our existing conservation deficit, creates a new conservation imperative. To ensure that species and ecosystems have the best chance to adapt to and survive climate change, we need to greatly increase the pace and effectiveness of existing conservation efforts and invent new ways of building resilience to changes that we cannot fully predict and for which we likely have no historical analogues. We do not know everything there is to know about future climate conditions and how species and ecosystems will respond. But we do know enough, following principles of conservation biology and increasingly clear projections of temperature and precipitation patterns, to take the much more aggressive steps that we are reasonably certain will be needed to prepare species and ecosystems for the coming changes.

Conservation biologists use the term *resilience* in thinking about how to prepare for climate change. Resilience refers to the ability of species and ecosystems to both resist changes and respond to changes and return to a stable state after some sort of shift. Since we cannot know precisely how the climate will change or how plants and animals will respond, we must look for ways to build resilience in natural systems. The size of natural areas or populations of animals, the condition of habitats, and the landscape context that defines how populations of species interact are all important for determining resilience. We must assess conservation efforts today and in the future by considering whether they increase the resilience of natural systems and species to future climate change.

To illustrate both the existing conservation deficit and the future challenges associated with the new conservation imperative, we can look to three major natural systems in the state of Colorado where conservation planning is fairly advanced. The three variables of size, condition, and context affect the likely resilience of these systems, and the species within them, in different ways, but illustrate the challenges they face.

The grassland systems of eastern Colorado are a sometimes overlooked piece of the conservation picture in the state, yet are indicative of one of the more pressing global challenges in the face of climate change. At a global level, grassland systems are the most threatened by past human actions, and the least protected against future changes. In the United States, nearly 50 percent of bird species that breed in grasslands are of conservation concern. In Colorado, more than 40 percent of our grassland habitats have already been converted to nonnatural land uses such as urban development and crop agriculture. Further, the remaining grasslands are increasingly found in smaller and smaller patches, with impacts impinging on them from surrounding land uses. A recent study by The Nature Conservancy found that only approximately 15 percent of the remaining native grasslands in eastern Colorado can be considered in a high state of integrity, as measured by the extent to which they are interspersed with roads, power lines, energy development, and other human land uses.

Projections of climate change for the eastern grasslands of Colorado indicate warmer and drier conditions, along with increasing likelihood of prolonged droughts. These sorts of changes will drive lowered productivity in natural systems and shifts in species composition that will affect the ability of habitats to support the current composition of species. Species that already survive in small patches of their once large habitat, such as the lesser prairie chicken, are increasingly at risk. To create resilience to these sorts of changes, the conservation imperative is to protect large areas of natural land, to avoid fragmentation of these areas, and to maintain connections between natural areas that permit movement of plants and animals as conditions change.

Given the sorts of responses needed to maintain resilience in the grasslands of Colorado, are current conservation efforts keeping pace? Unfortunately, it does not appear so. The grasslands of Colorado are mainly in private ownership. The dominant land use of much of these private lands has been cattle ranching. This land use has helped maintain the natural land cover and ecosystem condition that support native species and lend a level of resilience to changes in climate. However, native grasslands in private hands face a variety of threats that could reduce their resilience. Of most concern to date has been increasing encroachment of urban, suburban, and exurban development, including the spread of thirty-five-acre ranchettes, that reduces the size of large natural areas, increases the fragmentation of the landscape, and reduces the quality of remaining habitat for native species, particularly those most vulnerable. Our most effective tool for resisting this degradation of the grasslands and the loss of resilience to climate change has been conservation easements to keep native grasslands in large ranches and traditional land use. Unfortunately, of the 16.1 million acres of remaining natural lands in eastern Colorado, only 1.8 million acres, about 11 percent, have some form of permanent protection against conversion to land uses that will be less likely to sustain species and habitats in the face of climate change. To add to this fairly grim situation, a new threat to the grasslands of Colorado has arisen in the name of preventing the emissions that cause climate change: wind power and biofuels.

There is no question that severe climate change poses the greatest threat to the survival of species and ecosystems in the grasslands, but it would be a grave mistake to severely damage the resilience of native grasslands in the process of addressing climate change. Increased conversion of native grasslands to grow biofuels, and increased fragmentation of remaining intact lands by wind turbines that have been placed without adequate planning, will reduce the ability of grassland species and systems to adapt to changes in climate. The conservation imperative for the grasslands of Colorado, and all of the Great Plains of North America, is to protect the remaining natural lands and to plan wisely for the deployment of renewable-energy sources, so as to both reduce global-warming emissions and sustain the resilience to natural systems to climate change.

In the ponderosa pine forests of Colorado and the Rocky Mountain West, the conservation imperative is to improve the condition of the forests, to sustain biodiversity, and to build resilience to increased likelihood of uncharacteristic, catastrophic wildfires. Ponderosa pine forests are the most rich in biodiversity of Colorado's major forest systems. Because they occur at lower elevations and mainly along the Front Range, they are also the most heavily populated by humans. Ponderosa pine forests, particularly in lower elevations of the Rocky Mountains, developed in the presence of fire. Low-intensity, frequent fires killed younger trees and helped maintain an open canopy and a rich understory of grasses and forbs. The past century, with a combination of intensive logging, followed by grazing, followed by decades of fire suppression, has left much of the ponderosa pine forests densely stocked with trees. These forests are much more susceptible to intense crown fires that can kill all trees and damage forest soils to the point that recovery can take decades, even centuries. The risk to both the biodiversity of these forests and the people who live in them from catastrophic wildfires is greatly elevated from our past lack of stewardship.

With climate change, the risk to these forests and human communities is even greater. With just the trend in warmer and mostly drier conditions in forests across the western United States documented in the past two decades, the frequency, intensity, and size of wildfires have already increased significantly. In Colorado, the Hayman fire erupted in the unusually hot, dry summer of 2002, and caused nearly $40 million in damage, burned 133 homes, and forced the evacuation of 5,340 persons. Because conditions were so dry and the forest so dense, more than 138,000 acres of ponderosa pine forest were completely lost. Unlike with the fires these forests evolved with, recovery from the Hayman fire will take centuries. The combination of unhealthy forest conditions and the kind of climate conditions we face in the coming years has already taken a toll on Colorado's Front Range forests that will be felt for many generations. How many more fires like Hayman can Colorado endure?

Future projections of climate change for the lower-elevation forests of the eastern Rocky Mountains are for much warmer and drier conditions that will make wildfires even more likely and intense. Forests will be at increased risk of catastrophic loss. But perhaps much scarier than the projected changes in climate is the fact that so many of our forests head into this period of change in a condition that likely dooms them to severe ecological

damage if fire occurs. If we hope to sustain our native ponderosa pine forests in the face of even modest future climate change, we must drastically
increase the pace of restoration efforts. We know how to return our ponderosa pine forests to conditions that are more likely to resist catastrophic
wildfires that threaten our homes and the plants and animals that live in
them. It requires thinning of overly dense trees and judicious but greatly
expanded use of prescribed fire to return to and maintain the conditions
that sustained these forests for centuries. This is the conservation imperative
for these forests, and we are lagging far behind.

In most writing about climate-adaptation strategies in the arid West,
the emphasis is on water supplies. We live in a precarious balance between
competing human uses of water. It does not take much imagination to see
a future when some of those uses, whether for agriculture or future urban
growth, will lose the competition. The stresses of climate change and
reduced water availability make that outcome more likely. What often gets
lost in this debate is the need to keep sufficient water in our rivers and
streams to sustain the native species and all the benefits that properly func-
tioning freshwater systems provide us. The good news, so to speak, for our
rivers is that in many ways human development has already had as great or
greater an impact on these systems as we might expect from climate change.
In Colorado, we divert nearly 80 percent of natural flows from streams at
some point, and consume nearly half of the total. We trap peak flood flows
in reservoirs, and we take so much of the low summer flows in some places
that stretches of river run dry. If we can figure out how to sustain aquatic
habitats in the face of current human uses, then we will be taking some of
the steps to prepare for climate change. The bad news, of course, is that the
impact of climate change will come on top of the great changes we have
already wrought on our rivers and streams.

One of the clearest projections of future climate change is reduced average inflow of water to our rivers simply due to increased temperatures. In
addition, winter snowpacks will shrink, reducing natural storage and further reducing late-summer flows. The likelihood of severe drought will also
increase. If we do nothing to protect and restore flows, and we only pay
attention to competing human demands, we risk many more miles of dry
riverbeds, many more species at risk of extinction, and much more degraded
riparian habitats through expanded invasion of weeds like tamarisk. The
conservation imperative for rivers and streams is very clear: if we want to

have all the benefits that flowing water provides us in the West, we must consider the needs of nature in the decisions we make about water use. Protecting and restoring flows today, anticipating the additional stresses that climate change will present, are within our technical capacity. Whether we have the will to do so is less clear.

Many of us who live in the West consider ourselves deeply connected to our lands and waters. The natural beauty of our grasslands, forests, and rivers and the wildlife they sustain draw people to our region to live and visit. It is this natural world that many evoke when they warn of the perils of climate change. We, however, continue to cause impacts on the natural environment through the way we grow, the way we develop our energy supplies, the way we use water, and the way we spend public money today. These actions threaten the resources we cherish. They also create a conservation deficit that will make the impacts of climate change much worse, and which will make it much more difficult for us to adapt. We have a conservation imperative today, clearer in the Rocky Mountain West than in most places, that calls for us to begin the restoration and protection that climate change demands.

Tim Sullivan is director of conservation initiatives and acting state director for The Nature Conservancy in Colorado, based in Boulder. His academic background is in wildlife conservation biology, and he has worked on international, national, and state conservation policy initiatives for the past twenty-five years.

COUNTRY IN OVERDRIVE—
LAND USE, TRANSPORTATION, AND CLIMATE IN THE WEST
By Jocelyn Hittle and Ken Snyder

People who care about climate change, who take steps to reduce their carbon footprint by reducing energy use at home and in the office, and who wouldn't be caught dead in a gas-guzzler can nevertheless be guilty of contributing to climate change simply because of where they live.

In the United States, sprawling development, access to cheap or free parking and highways, the relatively low cost of fuel, and in many cases the lack of effective public transportation or bicycling and pedestrian options lead to personal vehicle use being the predominant form of transportation. Vehicle miles traveled (VMT) in the United States have grown by three times the rate of population growth since 1980.[1] Since the 1950s, new growth has been, and indeed often continues to be, designed around motor vehicles. As a result, many Americans spend more time commuting each year than they do on vacation.[2]

All this driving leads to significant greenhouse gas emissions. Over half of each household's emissions comes from their motor vehicles. Since the average American commuter produces six to nine tons of carbon dioxide per year, reducing the number of trips by car and the distance traveled will have a significant impact on emissions. If no action is taken to change the way Americans drive, emissions reductions from things like improved fuel efficiency or lower-carbon fuels will be completely nullified by increases in VMT.[3]

This clear link between VMT and climate change has led planners and researchers to ask how to most effectively reduce VMT. Studies indicate that a suite of urban design elements impact VMT, and that these elements are most effective in combination.

What Works

Because the question of how to reduce VMT has implications for many disciplines, over 100 studies have been completed to date on what design elements affect VMT. Taken together, these studies show that VMT reductions most often result from factors known as the five Ds: density, diversity of building/use types, destination accessibility, distance to transit, and design. Making destinations accessible by modes other than driving is the most important factor in predicting reductions in VMT. Distance to transit has the second largest effect on VMT, with diversity, design, and density roughly equal as the third most predictive elements. These elements predict and explain reductions in VMT even when self-selection (for example, residents specifically choosing to live in areas that provide alternative transportation) is taken into account.[4]

Several studies have determined that residents of more compact, diverse areas drive between 25 and 30 percent less than those in more sprawling areas. For example, residents in King County, Washington, who live in more walkable neighborhoods drive 26 percent fewer miles per day.[5] A meta-analysis of many of these types of studies shows that people living in places with twice the average density, diversity of uses, accessible destinations, and interconnected streets drive about 30 percent fewer miles, even when socio-economic status and other factors are taken into account.[6] This reduction in VMT suggests that emissions reductions of 7 to 10 percent from current levels could be achieved by 2050 through land-use changes alone.[7] By shifting 60 percent of new growth into more compact development patterns, estimates indicate, up to 79 million metric tons of carbon dioxide could be saved each year by 2030. This savings is equal to a 28 percent increase in federal fuel-efficiency standards and one-half of the cumulative savings of the new thirty-five–miles–per–gallon corporate average fuel economy standards.[8]

Areas that feature the right combination of the five Ds include many existing older neighborhoods, as well as newer mixed-use developments, transit-oriented development, or traditional neighborhood development. Increasingly, these types of development are given priority by municipalities because of high livability and corresponding benefits such as public health and reduction in obesity, and the developments' improved ability to fund regional amenities such as parks and transit.

Several western cities are looking to land-use and transportation planning as a way to meet their greenhouse gas–reduction goals. Large-scale transportation improvements, such as Denver's FasTracks regional rail and bus system, are a high-profile step toward reducing VMT while reducing commute times for a traditionally car-oriented city. Development around new light-rail stations in particular is focusing on the Ds. For example, the city is visualizing station areas as dense, mixed-use destinations in themselves. Similarly, Albuquerque has made a concerted effort to increase public transit ridership and is working to connect their existing bicycle paths into a cohesive system to reduce bicycle commute times.

To more directly address the land-use component of climate change, Salt Lake City is undergoing a revision of its zoning and subdivision codes with a focus on sustainability goals, including climate change. Some of the recommended changes to their code include areas of mandatory mixed use and/or higher density and removing regulations that discourage green renovations of existing structures.[9] The city's goal is to reduce emissions, via these and other methods, by 3 percent per year for the next ten years, with a long-term goal of reducing emissions 70 percent by 2040. Salt Lake City's work, when complete, ideally will be an example to other communities of ways to concretely link land-use decision making to its climate implications.

A Tale of Two Cities

As a more detailed example, consider the towns of Houten, in the Netherlands, and Milton Keynes, in the United Kingdom, two cities that have made sustainable development a primary goal of city policy. Contributing author Ken Snyder studied these two cities as part of a German Marshall Fund fellowship. Both cities were designed with heavy bicycle use in mind. Milton Keynes has a cycleway system of bicycle paths separate from the street grid. These paths follow rivers and wind through natural areas. As a result of this extensive bike trail system, Milton Keynes has more bridges than Venice. In spite of this, Milton Keynes's ridership is only 7 percent, an improvement over the United Kingdom's national average of 2 percent. By comparison, Houten has a system of sixteen car-tight sectors—each sector is only accessible by car via one entrance from the ring road. Although bicycles are not allowed on the ring road, both bicycles and pedestrians are given priority on the woonerfs, the interior roads, where most of the housing faces. This system creates little vehicle traffic and has boosted Houten's bicycle ridership to 70 percent. (The national average is already high at 30 percent.)

Why the difference? Cultural differences between the United Kingdom and the Netherlands obviously play a significant role. Yet, ultimately, Milton Keynes's cycleways do not provide enough convenience and connectivity between homes and jobs or other common destinations. The trails often take more scenic but longer routes, and no steps were taken to make driving less convenient. The system's design means that driving remains the most common mode of transportation. Milton Keynes's roads are wide and designed for efficient movement of vehicles, and cheap or free parking is readily available throughout town.

By comparison, in Houten if you wish to travel half a mile to a nearby restaurant in the adjacent car zone, the short and direct trip by bicycle is much more complicated in a car. By car you have to travel outward to the ring road surrounding town and then reenter in the adjacent car zone. If people mainly drove this road system by car, VMT would dramatically increase, but because driving is significantly more inconvenient, travel by bike is the predominant mode.

These two cities illustrate that unless driving is more expensive and/or less convenient, improvements in alternative transportation will reduce VMT, but perhaps not significantly enough to help communities meet their ambitious emissions-reduction goals. Instead, more comprehensive urban design, using several strategies working in concert, is required to significantly reduce greenhouse gas emissions.

Designing and Measuring

Because reducing VMT takes a multifaceted approach, communities interested in effective design need to employ a variety of design tools. A set of tools has evolved that help communities measure where they have been, understand where they are, and look forward to where they want to be. These tools, often referred to as decision-support tools, can be used to model a new development's potential impact on VMT and emissions, help communities visualize and analyze different development scenarios, engage the public, and measure progress toward goals.

For example, communities interested in creating a development that reduces VMT could employ a spatial tool that measures and compares indicators, in particular the five Ds, among scenarios. For example, a tool could measure street connectivity, average distance between residences and amenities, overall mix of uses, and the balance of jobs and housing. There are several tools that provide this type of analysis, including Placeways, LLC, CommunityViz tool, and Criterion Planners' INDEX tool.

These tools, and others like them, also allow citizens in communities to imagine what the future of their community might look like using two- and three-dimensional visualization. Next to outdated regulations, one of the biggest barriers to designing and implementing new community types is public resistance. Using tools and photographs to help citizens "see" what neighborhoods will look like can be key in overcoming this resistance to development that is accessible, walkable, dense, and diverse enough to have an impact on VMT. There are a growing number of examples of developments that reduce VMT and are practically and visually appealing to even die-hard density opponents. Using analysis tools and highlighting these appealing developments as examples help reduce community fears of density and mixed-use developments and are vital to designing for greenhouse gas reduction.

Communities looking to meet emissions-reduction goals need to measure the effects of current design and future development scenarios on greenhouse gas emissions. Many decision-support tools, including INDEX and MetroQuest, have added the capability to directly model VMT and greenhouse gas emissions associated with various development scenarios. Spatial analysis tools and transportation modeling tools increasingly provide greenhouse gas emissions as an output. Packaged tools for measuring a community's sustainability overall, including VMT and greenhouse gas reductions, are in development. A list of land-use planning and community-development tools that can help communities reduce their VMT and thus carbon footprint can be found at www.planningcollaborative.net.

Tools that easily and accurately synthesize public input in the planning process are equally important to helping communities understand and choose new development types. These tools include online polling, mapping, forums, wikis (collaborative document editing), and even text messaging. These tools can be used to inform and educate citizens about the challenges their communities face, the effects of various development scenarios on indicators, including VMT and greenhouse gas emissions, and progress toward implementing strategies to meet community sustainability and greenhouse gas reduction goals.

Progress toward community goals can also be measured using decision-support tools over time. Flexible, results-oriented planning allows for regular revisiting of plans and tailoring of new plans and developments to meet communities' changing goals and needs. As communities adopt plans that aim to reduce VMT and greenhouse gas emissions, these tools will be important for measuring progress against benchmarks. As national attention turns toward regulating greenhouse gas emissions, in particular carbon dioxide, as a pollutant, the ability to repeatedly and accurately measure current conditions and future scenarios with respect to carbon dioxide emissions becomes ever more necessary.

For example, Redmond, Washington, and Boulder, Colorado, have been using various tools to evaluate how their multimodal travel is helping meet their climate-action-plan objectives. Redmond has developed a Mobility Report Card that monitors the performance of their Transportation Master Plan and their multimodal travel objectives. Likewise, the City of Boulder has been tracking the progress of their Transportation Master Plan's ability to shift travel behavior to achieve sustainability objectives for almost two decades. The city measures travel behavior and evaluates the success of their investments in multimodal infrastructure.

In addition to using tools to keep track of greenhouse gas emissions reductions, communities can use decision-support tools to determine and demonstrate to stakeholders that they will save money if they implement urban design that reduces emissions. In particular, savings accrue if policies are put into place as new growth happens rather than after the fact.

The Opportunity

According to research conducted at Virginia Tech, by 2050, new construction in the United States will include 89 million new or replaced homes and 190 billion square feet of new offices, institutions, stores, and other nonresidential buildings. If forecasts are correct, two-thirds of the existing development by 2050 will have been built after 2007.[10] The siting and design of those developments will be determined in the coming years and could go a long way toward reducing emissions from transportation.

In a fortunate confluence of the private market and public benefit, communities that offer choices in terms of housing, those that offer mixed-use living, and those with shorter commute times to employment are ever more appealing to consumers. Studies conducted by real estate researchers and universities have found that about one-third of all homebuyers prefer "smart growth–style" communities. For the first time, prices of attached units are higher than for detached single-family dwellings. The Brookings Institution also has discovered that because demand outstrips supply, the price premiums on homes in mixed-use developments are 40 to 100 percent.[11]

If two-thirds of the 2050 built environment has yet to be built, and if people are eager to buy or rent the type of homes, offices, and industrial properties that help reduce VMT and greenhouse gas emissions, we face a tremendous opportunity to design this new growth with climate change in mind.

What's Next?

We are poised to take advantage of the growth that the country, and the West in particular, faces in the next thirty years. Decision-support tools will help communities make improved land-use decisions based on both livability factors and long-term climate goals.

The time is right to shift the focus of community and urban design policy and practice to reflect all that we now know and continue to learn about the links between land use and climate change. There is an opportunity to make the link between planning and greenhouse gas reduction an explicit basis for decision making. The choices that individuals make about where they live are already shifting to be more climate friendly. With communities continuing to encourage or require new growth, citizens will have every opportunity to leave the car at home and be truly climate conscious.

Jocelyn Hittle is the director of planning solutions for PlaceMatters, a nonprofit organization that promotes environmental, economic, and social sustainability in decision-making processes. She focuses on holistic planning processes, including linking land-use planning to ecosystem science. Until recently, she also was the editor of *Planning & Technology Today*, the publication of the American Planning Association Technology Division. She is a graduate of Princeton University and Yale University's School of Forestry and Environmental Studies.

Ken Snyder is president and chief executive officer of PlaceMatters. He is a nationally recognized expert on a broad range of technical and nontechnical tools for community design and decision making. He is a graduate of Oberlin College and Yale University's School of Forestry and Environmental Studies.

Notes

1. Reid Ewing, Keith Bartholomew, Steve Winkelman, Jerry Walters, and Don Chen, *Growing Cooler: The Evidence on Urban Development and Climate Change* (Washington, DC: The Urban Land Institute, 2008).
2. Ibid.
3. Ibid.
4. Ibid.
5. Lawrence Frank and Company, 2005. "A Study of Land Use, Transportation, Air Quality, and Health in King County, WA." (In collaboration with The Neighborhood Quality of Life Study, Center for Clean Air Policy, Geostats, LLP, and McCann Consulting).

6. Reid Ewing, Rolf Pendall, and Don Chen, 2002. "Measuring Sprawl and Its Impact."
7. Ewing et al., 2008.
8. Ibid.
9. "Public Review Draft of the Sustainable Code Diagnosis," Salt Lake City, February 23, 2009.
10. Ewing et al., 2008.
11. Ibid.

PART THREE
COMMERCE AND INDUSTRY

A recent *Newsweek* article declared Denver to be the green-jobs center of the country—ground zero for the birth of the New Energy Economy—due to the growth of new solar and wind firms, even in the face of an otherwise struggling market. And while the "old energy economy" (the one based on oil, gas, and other extractive industries) has been the undisputed basis of past growth and development in this city and region, the new trend may signal the future not only for the Rocky Mountain region, but for the world.

New Energy Economy, "green-collar" jobs, carbon markets—all are buzzwords familiar in the intermountain West and gaining currency at the national level, in President Barack Obama's stimulus package and elsewhere. The writers in this section explore the business of global warming and related trends in our region—how the current "green rush" has both been shaped by public incentive and driven by old-fashioned entrepreneurial drive and opportunity.

Energy analyst James R. Udall's pieces address two ends of the new energy spectrum—an ideal scenario (the Danish island of Samso) and the less idyllic reality in Colorado today (a case study from the staff of Holy Cross Energy attempting to meet the Colorado governor's long-term climate goals). Boulder journalist Todd Neff addresses American energy habits from a more personal perspective and describes how our habits drive the demand for new supplies. Businesswomen Catherine Greener and Martha Records tell how regional

entrepreneurs are advancing new clean-technology and green-business models. Green-building and sustainability consultant Josh Radoff describes the challenges of improving the way we design and construct our environments. Matthew H. Brown describes a variety of new mechanisms that have emerged in both public and private sectors to finance energy efficiency and renewable-energy investments. And Susan Innis, administrator of the state-sponsored Colorado Carbon Fund, describes how state government has engaged in the market for carbon offsets as a possible way to provide funding for local renewable energy and other types of carbon-offset projects.

Extreme and historically unusual weather events appear to be as much a signal of climate change as any single drought or heat wave. Insurance companies became activists with regard to global warming earlier than other industries as they observed extreme weather risks and exposure levels rising internationally. Aspen Skiing Company has famously described the "death of snow" as a motivator in its own climate activism, but in the summer of 2009 the weather was more noticeably wacky than warm in Colorado, even as the Pacific Northwest experienced record-breaking heat and Texas experienced the worst drought in fifty years. The past decade along the Colorado Front Range has included some of the driest and wettest years on record, with related insurance losses at both ends. Another type of risk analysis for business extends to understanding the variability of energy costs, as the price of oil varies widely from month to month and year to year. The business case for development of the New Energy Economy, in addition to reducing the risk of climate change–related impacts on Earth's natural systems, includes the potential to stabilize energy supplies and costs through the increased reliability and stability of renewable-energy sources.

Finally, this section includes two perspectives on the future of conventional energy sources: Michael L. Beatty describes the potential for natural gas to serve as a "bridge fuel," while Steve Andrews posits a more conflicted scenario of the world's energy future under peak-oil projections.

THE BIG BONFIRE—
WHAT COLORADO CAN LEARN FROM THE SAMSO EXPERIMENT
By James R. Udall

President Barack Obama speaks about the "tyranny of petroleum"—but can we really turn our backs on fossil fuels, the black magic that invented prosperity? Is not oil more addictive than crack cocaine? Do you know anyone who has given it up, gone cold turkey? Well, there is one place where people have begun to kick the habit, a tiny island off the coast of Denmark. What does their remarkable experience have to teach us?

Last fall, a Viking washed up on my doorstep. His name was Soren Hermansen. For the past ten years, he has spearheaded one of the most audacious experiments in the world: the attempt of 4,000 people living on the small Danish island of Samso to liberate themselves from fossil fuel.

A few weeks after being anointed an environmental hero by *Time*, Hermansen came to America on a speaking tour with his wife, Melane, an artist and photographer.

Everyone wants to hear the story of Samso. In 1997, Denmark held a contest to select an island that would eventually be run entirely on renewable energy. Samso won. In the decade since, the islanders have invested $70 million of their savings and government grants in wind turbines, district heating plants, solar panels, and biofuels. Today, they are energy independent. Their carbon footprint is not just small, it is negative, since they produce more energy than they consume.

Reporters who visit the island sometimes describe its farmers as "beefy." Like farmers everywhere, those on Samso occasionally have difficulty finding wives. This led them to create a website called farmerdating.dk. The personals are in Danish, but a typical one reads: "Beefy farmer with large tractor seeks attractive woman with boat. Must be able to sew and clean fish. Send picture of boat and motor."

Hermansen grew up on the island raising beets and onions. One day, Melane, an urban refugee with dyed crimson hair, arrived. She needed a place to stay; he had a room to rent. They are an affectionate pair. Drinking wine one night, I asked them about their initial attraction.

"Soren looked like he might have a big tractor," she said. "Melane had nice jeans," he recalled, "and seemed perhaps suitable for breeding."

In Telluride, Colorado, Hermansen showed slides to an overflow crowd. Tousled by the incessant wind, the ruddy-cheeked natives of Samso are conservative, rural, and intensely pragmatic. Since World War II, they have imported coal-fired electricity through an underwater cable from the mainland. Fuel for their tractors and automobiles was delivered, expensively, by ship.

When the islanders first learned that Samso had been selected to be Denmark's "renewable-energy island," many were skeptical. A proposal to build a centralized district heating plant that would provide heat and hot water to hundreds of homes was finally approved—but only after many meetings. As the years passed, the islanders began to embrace renewable energy as a business opportunity, a brand, an ethic, something akin to sport. Pensioners insulated their homes. Teachers installed solar systems. Their ambitions grew, and under Hermansen's leadership, they successfully raised $40 million to construct an offshore wind farm, ten gleaming white towers hovering over the blue sea.

Today, the farmers of Samso seine the sky, shipping a surplus of clean power to the mainland. On calm days, they import. During the summer, the turbines lure thousands of tourists to the island. "They spend the night and their money, we shear them like sheep," Hermansen said.

What does any of this have to do with the West? Everywhere I went with Hermansen, we heard that "America isn't Denmark." Fair enough. The Danes pay three times as much for electricity as we do (and use 40 percent less). They pay $6 for gasoline. In short, there is no Powder River basin in Denmark, no place to mine a million tons of coal each day. (Wyoming is sometimes described as an "energy state," but it is more akin to an "energy nation," since only four countries produce more energy.)

The irony is that the West's renewable resources are better than Denmark's. Far better. In many parts of the Great Plains, each square mile gets swept by $5 million worth of untapped wind power per year. The solar flux raining down on the desert southwest is worth $2 million per square mile per year. Within ten miles of Medicine Bow, Wyoming, you could plant enough wind turbines to run the entire state. At noon on a sunny day, there's fifty horsepower of sunlight striking your south-facing roof. But drowning in fossil fuels, we turn our back to the sun. Coal stymies wind. Natural gas blocks biomass.

Outdated policies are part of the problem. In Europe, the grid is increasingly a two-way street, with easy access, transparent rules, and attractive tariffs that reward farmers and homeowners for producing power. Here in the West, however, the grid is like a highway whose on-ramps are blocked with "Do Not Enter" signs, stifling innovation and independent power providers. These differences help to explain why 5 percent of Danes own shares in utility wind turbines, while most Americans have difficulty imagining themselves as energy producers rather than mere consumers.

Of course, in a crisis, people and policies can change. In 2008, $17 billion worth of wind was installed in the United States. Many legislatures have adopted mandatory renewable-energy standards. Rural electric utilities promote geothermal heating, solar, and small-scale hydropower. Rifle and Eagle, Colorado, have announced plans for multimillion-dollar solar farms. A municipal utility in Lamar, Colorado, operates its own wind farm.

Speaking in Telluride, Hermansen told the schoolchildren: "Don't fret about the polar bear. Don't think global and act local. Just act local. If enough of us do, then someday we might do something good for a polar bear."

Worldwide, humans now burn about 1 million tons of oil, coal, and natural gas each hour. Given the size of this bonfire, is Samso just a fairy tale or is it possibly a harbinger? Is it a singular anomaly or a practical analogy, something the United States might someday replicate? After all, there are roughly 75,000 times more Americans than Samsonites. Could you run the United States on flows of carbon-free energy, or is that simply an eco-chic pipedream, green blow?

Thinking about it, it is not particularly difficult to imagine running Nebraska (population 1.8 million) or South Dakota (800,000) on corn ethanol, hydropower, and wind energy. Both states are in a way just lightly populated solar collectors. But what about Atlanta or Phoenix or Los Angeles, where the air conditioner comes on in April and stays on through September, and the freeways are clogged at all hours? Americans aren't Danes, content to live modestly, close to home; our appetite for energy and mobility is prodigious, godlike. Indeed, each of us consumes as much energy as a small pod of killer whales or a 65,000-pound orangutan. Live free or die, we now use the energy equivalent of 100 pounds of coal or eight gallons of gasoline or 1,000 cubic feet of natural gas or one lightning bolt worth of energy per person per day. Is it any wonder that we lead the world in carbon emissions?

From Obama on down, we are the Oil Tribe. The defining ritual of our civilization is not Monday Night Football or church on Sunday, but buying a tank of gasoline. It happens 150 million times each week. Granted, there's a strange disconnect in this relationship: how can petroleum matter so much and mean so little? Muslims bow to Mecca five times a day, but in a country where mobility is our religion, we don't seem to care that oil won't be here forever. Gas pumps are as central to our way of life as bison were to the Sioux, yet while the Native Americans celebrated the life-giving animal in dance and ceremony, after filling up at the Kum & Go all we do is whine about the cost.

The largest faith-based religion in America is that we will never run out. I once saw a bumper sticker on a drilling rig that captured that sentiment: "Earth First, We Will Drill the Other Planets Later." It reminds me of a backcountry ski trip I took some decades ago with my brother, Brad. After skiing 100 miles, we arrived at a food cache we had placed weeks earlier. We were both famished. After wearily digging up the food, Brad grabbed a box of Mystic Mints, his cookie ration for the next 100 miles, and hammered down every single one in a few short minutes. I shot him an "I can't believe you just did that" look. "Easy come, easy go," he laughed.

Prominent energy historian Vaclav Smil argues that "our transition away from fossil fuels will take decades—if it happens at all." History teaches that energy transitions occur slowly, at a near-glacial pace, and for that reason Smil believes that Al Gore's proposal to repower America, so that we produce 100 percent of our electricity from renewable energy within ten years, is lunacy: "To think that the US can install in ten years wind- and solar-generating capacity equivalent to that of thermal power plants that took nearly sixty years to construct is delusional." Don't sugarcoat your diagnosis, Dr. Smil. "Gore has succumbed to what I call Moore's curse," the belief that dramatic improvements in computing power can happen within energy systems.

I sympathize with Gore. I wrote my first article on climate change some twenty years ago, and in the interim I have learned that obsessing about climate change can drive you insane. It's not that the crisis isn't real—it's that the scale of the necessary response is, well, almost unthinkable. What atmospheric chemistry requires, politics and inertia prohibit.

Google has a 2030 energy plan in which renewable energy becomes cheaper than coal. For this forecast to come true, the cost of producing solar power would have to fall about 300 percent. Nobel chemist Richard Smalley (of nanotechnology and buckyball fame) spent the last few years of his life investigating the energy dilemma. Solving the problem, he concluded, "would take ten miracles, in energy storage, photovoltaics, nanotechnology, superconductivity, and related fields. I'm a scientist and I think we might find those miracles if we went looking for them." But we haven't been looking for them, at least until recently. No bucks, no Buck Rogers.

Even if we were investing in RD&D, energy is not IT, Smil argues; it is an infrastructure play, not software and tech. It's also true that concentrated hydrocarbons (coal, oil, natural gas) are different from distributed energy flows (wind, water, and sunlight). Nature did 100 million years' worth of work on the former; they can be stored on-site and burned at will. Flows, on the other hand, are carbon-free, but they come as they are and must be captured before being converted. Because these fluxes are dilute, capturing significant quantities requires large nets. Think Three Gorges Dam or 180-mile-long Lake Powell.

The easiest way to solve the climate problem is to become innumerate. Owning a calculator makes things more difficult. According to Princeton scientists Rob Socolow and Stephen Pacala, keeping a billion tons of carbon out of the atmosphere each year by 2050 requires a forty-fold expansion of global wind power (plant turbines like trees) or a 700-fold expansion of solar electricity or building a new nuclear plant every three weeks for the next forty years. To solve the climate problem, we need many of these "wedges," not just one. We would need to wean Americans off oil and replace the coal that 3 million Chinese went underground today to mine. (One hundred will be dead by week's end.) In light of such numbers, Samso seems like a fairy tale and the climate challenge seems hopeless. Good-bye, Arctic ice; sayonara, polar bear!

If those are the technology hurdles, the political ones seem equally daunting. Bill Moyers has written that "one of the biggest changes in politics in my lifetime is that the delusional is no longer marginal." The well-funded climate-denial campaign is a great example. Surveys show that the more educated a Republican is, the *less* likely he or she is to believe in global warming. Indeed, a Republican is about as likely to believe that the sun revolves around the Earth as that humans have anything to do with planet warming. Reaganomics was based on the principle that rich and poor would get the same amount of ice, but that the poor would get all theirs in the winter. We are way past that now. But even coal-state Democrats aren't eager to cap carbon, and with good reason, since it will be expensive. One social scientist wrote that "it's hard to get the political animal to focus on a 100-year problem when the animal will certainly be dead by then and the machine may likely be broken."

The West has already warmed 2°F, and there's another 2° "in the pipeline" due to gases already emitted and the inertia Smil describes. For any additional increase in temperature, it's possible to calculate approximately how much fuel can be burned in the interim. In other words, we can derive a global carbon budget for the next 200 years. The diplomatic challenge is figuring out how to divide that budget among nations. The ethical challenge is to divide it among generations. Most current proposals short the grandkids, because reducing emissions will be costly and what have they ever done for us, anyway? Of course, *not* reducing emissions will also be expensive—we are busily creating dozens of new New Orleanses, cities doomed to drown—but someone else, distant in time and space, will have to pick up the tab for that. With climate costs, as with the Iraq war, the political imperative is to FedEx the bill forward.

━━━━━━━━━━━━━━━━━━

Around 1760, James Watt invented the coal-fired steam engine. In the centuries prior, coal had been used to smelt metal and heat buildings. But Watt's genius was to figure out how to turn heat into motion. The first use of his new engine was to dewater British coal mines, and over the next century coal and steam and iron learned to feed on each. It was not until the 1880s, however, that fossil fuels supplied more global energy than wood, water, and wind. In effect, the good folks on Samso are simply trying to use modern technology to turn back the clock, to stuff the genie back in the bottle.

According to the Greeks, when Pandora opened her box and released Evil into the world, she snapped it shut again, trapping Hope inside. In hopeful moments, I can envision a scenario in which America rouses like Rip Van Winkle from its sport-utility-vehicle slumber to tackle the climate challenge. Smil is right that in normal times energy transitions happen slowly. But this is not a normal time.

The first thing we need to do is change the frame. I like powder skiing, pikas, and polar bears, but it's a reach to expect the average Joe to give a damn about the climate or about global warming. Prosperity is another story; preserving it is the most durable and appealing narrative in American politics. An irrefutable case can be made that current energy policies are cannibalizing wealth and well-being.

One of the erroneous beliefs of the Intergovernmental Panel on Climate Change is that fossil fuels are superabundant, that we have literally centuries of coal, oil, and natural gas. This enables them (as it does Gore) to fixate on the bonfire's smoke, the exhaust, the emissions, and to ignore the fire. But by 2012, it will be clear to everyone that global oil production peaked in the summer of 2008. The dangerous belief that climate change is our most pressing energy problem will be revealed as an illusion. When it becomes indisputably clear that we are playing for more than we can afford to lose, both political parties will recognize that the empire is at risk, and that we'd better figure out how to leave oil before it leaves us.

The current recession has hit the reset button, and no one can be sure how soon the American economy will reboot. But by 2011, or whenever the global economy tries to grow again, a new oil price shock is guaranteed. As peak oil pushes climate change off the front page, there will be a reflexive, last-ditch effort to jump-start unconventional liquid fuels, coal-to-liquids, and oil shale. But these solutions are a mirage—dirty, slow, and costly, with abysmal energy return. Most everyone already understands that the future lies in electrifying transportation and in making green the new red, white, and blue. In short, reducing carbon emissions is likely to be a by-product of, not the driver in, enormous investments in energy security.

The most memorable line in President Obama's first inaugural speech was that "everywhere we look, there is work to be done." The citizens of Samso spent about $15,000 to secure their energy future, or about $1,500 per person per year over a decade. This is almost exactly what we Americans spend on the defense budget (and much, much less than we spent in 2008–2009 to rescue failed banks and bankers).

Garrisoning troops in 140 countries to preserve Yankee access to the world's dwindling store of oil is a bankrupt strategy. With unemployment likely to be near 10 percent, we are going to need jobs, and the most lucrative thing we can do is to deploy an army to stop the bleeding of energy. Efficiency and conservation will be the clarion call, with the fate of the economy hanging in the balance. This trying moment, now clearly and unmistakably visible on the horizon, will be a national test, a wake-up call, a determination of whether we still have the right stuff.

This retooling has been delayed far too long, which suggests we face a turbulent period. The good news is that for the first time in our history we might begin to grasp what the English poet William Blake meant when he wrote that "energy is eternal delight." By 2020, Americans may come to understand how 4,000 Danes living on a tiny island in the North Sea could think of energy as akin to sport, take pride in producing it, and consume it with utmost frugality.

In that sense, yes, America's best days lie ahead.

James R. (Randy) Udall developed Colorado's first solar-energy incentive program, the world's first renewable-energy mitigation program, and some of the most progressive green-power purchasing programs in the country. Udall is cofounder of the Association for the Study of Peak Oil and Gas–USA and writes from Carbondale, Colorado.

RUNNING DOWN AN UP ESCALATOR—
REDUCING COLORADO'S ELECTRIC-SECTOR CARBON EMISSIONS
The Staff of Holy Cross Energy and Energy Analyst James R. Udall

In Colorado, as in China, the petroleum-fueled automobile and coal-fired power plant are at the center of the climate challenge. To slow and then halt global warming will require dramatic changes in how we produce and use electricity, and equally dramatic changes in how we transport people and move goods.

In 2007, Colorado governor Bill Ritter asked the state's electric utilities to develop a plan to reduce their carbon dioxide emissions 20 percent by 2020 and 80 percent by 2050 from their 2005 levels. In response, Holy Cross Energy, a rural electric utility, studied how it—and the Colorado electric sector as a whole—could meet those ambitious goals.* Here is a snapshot of our findings:

- The growth dynamic—the difficulty of running down an up escalator—complicates efforts to reduce emissions. Cutting greenhouse gases in the face of increasing demand for electricity will be difficult and potentially expensive.
- In 2005, coal plants produced 72 percent of Colorado's power, natural-gas plants 24 percent, and hydropower, wind, and solar the rest. To meet Governor Ritter's 2020 goal, coal use would have to fall by nearly half, while natural-gas generation would have to nearly double. Carbon-free resources and efficiency savings would need to provide one-third of the state's electricity by 2020, up from 4 percent in 2005.
- Up to 2,700 megawatts of the state's existing coal generation, out of about 4,900 megawatts operating today, would need to be retired and replaced

---

* A few caveats are necessary. The Holy Cross Energy analysis did not calculate the cost of inaction, of the "negative externalities" that could be caused by a warming climate. It also did not attempt to calculate either positive or negative effects on the Colorado economy due to climate action, nor does it try to predict the timing or form of federal climate policy.

with combined-cycle natural-gas plants, wind farms, and other lower-emitting generation technologies.

- The statewide price tag for meeting the reduction targets would rise over time as natural gas replaced coal and new, lower-emitting plants were built. If electricity consumption grows by 2 percent annually, then by 2020, Coloradans could be spending $1 billion more each year for electric power, above and beyond normal price escalation.
- If Colorado wants to reduce its carbon footprint, bipartisan leadership will be crucial. Citizens must be engaged in the discussion, since effective climate policies can only be crafted with their support, and because they will have to pay the costs of climate protection.

The Growth Dilemma

When it comes to emission reductions, growth is often ignored. This is unfortunate, because growth is a big deal. If demand for electricity were not growing, Holy Cross Energy's emissions would be much lower today than they were a decade ago, because of changes in our fuel mix. In the same vein, absent population growth the United States would be on track to meet the emission targets established by the Kyoto Protocol. Dr. Al Bartlett, professor emeritus of physics at the University of Colorado, believes that the "greatest shortcoming of the human race is our inability to understand the implications of exponential growth." Bartlett goes on to explain how seemingly trivial growth rates can have momentous implications. The American electricity industry is a classic example of exponential growth. Between 1949 and 2000, demand for electricity grew by 5 percent per year. By 2000, the nation was using thirteen times more electricity than it had at the end of World War II.

In recent years, Colorado electricity demand has been driven by economic activity, population growth, and the proliferation of flat-screen televisions, air conditioners, microwaves, computers, cell-phone chargers, and other electrical appliances. In 2005, a typical Colorado household used nearly 900 kilowatt-hours more than it did in 1995. This does not make it impossible to reduce emissions, but it does make it tough sledding.

In a recent report, the Colorado Energy Forum forecast that statewide electricity demand is likely to grow an additional 40 percent between 2005 and 2020. The implications are sobering. To achieve a 20 percent reduction in Colorado's electric-sector carbon emissions and to simultaneously meet soaring demand, the state's utilities would need to cut their "carbon intensity"—the amount of carbon dioxide produced per unit of electricity—almost in half.

A Herculean Task

At the end of the day, reducing electric-sector emissions means burning less fuel, burning cleaner fuel, or using fuel more efficiently. It is also possible to use flowing water, wind, and sunlight (or nuclear power) to displace electricity now provided by fossil fuel. If the state's population were not growing, Coloradans could achieve the governor's goal by burning 20 percent less oil, natural gas, and coal. This would be a significant undertaking, since everyone would have to participate, but it's not impossible. In the electric sector, half those fuel savings might be gained through conservation programs and energy-efficiency investments. The other half could be obtained through improvements in carbon intensity. For example, utilities could displace some coal-fired electricity with natural gas, which has about half of coal's carbon content. Or they could build wind farms, burn beetle-killed trees in biomass plants, install solar systems, and so forth.

But, of course, Colorado's population is growing. Indeed, forecasts are that the state might see 1 million new residents by 2020. Now, in order to reduce emissions by 20 percent, someone needs to convince the newcomers not to use any fuel to heat their homes, run their cars, or power their businesses. If, as seems likely, they insist on burning their share, then improvements in energy efficiency and carbon intensity would have to occur on a herculean scale.

Burn Less Coal?

If electricity consumption continues to grow, the only feasible way to reduce Colorado electric-sector emissions is to burn much less coal. To obtain a 20 percent reduction, Colorado's fuel mix would have to change dramatically by 2020. It is difficult to communicate what an enormous transformation it would be to substitute gas, conservation, and renewables for coal as demand expands. To sum up: as electricity production grows by 38 percent, coal use falls to only 30 percent of the power supply, natural-gas use nearly doubles to 40 percent of electric supply, and carbon-free generation and efficiency grow rapidly to provide 30 percent of electricity services—all in a bit more than a decade. This would be the largest, and most costly, change in the history of the Colorado electric industry.

Many Critical Questions

Before undertaking such a dramatic course change, many questions must be addressed. Is such a rapid transformation physically possible in the time allotted? Since coal plants provide the state's cheapest power, what would the "dash to gas" cost? (About $500 million per year by 2020 in increased fuel costs alone.) How many new power plants would need to be built, in what locations, when, and by whom? (A complete analysis remains to be done, but many new plants would be needed.) Are there operational constraints to bringing on large amounts of new wind? (Yes, some.) Are existing gas pipelines adequate to the task? (No, upgrades would be needed.) Would contractual obligations inhibit this scenario? (Yes.) Would the economic impacts of this switch affect all utilities equally? (No.) How much higher would bills be at the end of the period? (Significantly.) Would the added costs bring clear benefits? (Possibly, if activities in Colorado were coordinated with those in other states and nations.) What happens after 2020? (Solving the climate problem would require large additional reductions here and overseas.) Is there an overlap between actions in the electric sector and reduction opportunities in the transportation sector? (Yes, think plug-in hybrid and all-electric vehicles.) Could conservation reduce the number of new power plants and fuel switching required? (Clearly.) Finally, the most critical question of all: do electric-utility customers, citizens, and voters support making the large investments that would be necessary? In a recent poll, a healthy majority of Holy Cross Energy consumers were willing to pay 5 to 10 percent more to address climate change, energy security, and sustainability. On the other hand, Amendment 37, the passage of which in 2004 established mandatory renewable-energy standards, was only narrowly supported by Colorado voters, even though it specifically limited electric-bill rate impacts to 1 percent.

The Cost of the Power Switch

What would it cost to meet Governor Ritter's reduction goal? In rough numbers, our analysis suggested that the price tag would be about $1 billion per year by 2020. The increases fall into two categories: costs associated with substituting natural gas for cheaper coal and costs associated with building new power plants.

Colorado's natural-gas production has exploded since 1990. The state now produces about 7 percent of the nation's supply, most of which is exported. Depending on the future price of natural gas, the annual cost of diverting one-fifth of the exports to in-state power production could be $500 million or more by 2020.

Colorado's emissions of carbon dioxide have been growing for more than a century. To put them on a different trajectory, existing coal plants will need to be replaced with lower-emitting generators. This is a costly proposition, particularly under a tight deadline. That's because the difference between the depreciated value of an existing power plant and the cost of a new one is substantial, even staggering. Today, many of the older coal plants in Colorado have an amortized value of about $300 per kilowatt of capacity. New plants—of any kind—will cost three to twenty times more, an investment that would need to be recovered through significant increases in electric bills. An automotive analogy may be useful: continuing to drive an old beater is much cheaper than buying a Toyota Prius.

Since new power plants can cost a billion dollars or more, they are not bought on a whim. Once built, they operate for decades. As their capital costs are recovered, well-maintained plants become sources of inexpensive power. This explains why many of the nation's coal plants are thirty, forty, even fifty years old. The state—and the nation—face a dilemma: from an emissions standpoint, we can't afford to continue operating all existing coal plants, even though some of their capital costs have not yet been fully recovered. But from a fiscal standpoint, can we afford not to? Is there a way to mitigate, if only partially, this conflict? Yes, by extending the proposed 2020 deadline some years, and by investing heavily in energy conservation, it is possible to reduce the cost of achieving our reduction goals.

The Role of Energy Efficiency

To be successful, climate policy must be enduring. To be enduring, it must be cost-effective. The cheapest ways of reducing emissions are found in the realm of energy efficiency and conservation. Effective statewide efficiency programs that reduced growth in electricity consumption could save billions in avoided power plant costs. Although Holy Cross Energy has promoted efficiency for a decade, we have learned that our ability to control what happens on the customer's side of the meter is limited. We can provide advice, encouragement, free programmable thermostats, and solar rebates, but if a homeowner wants to add a second refrigerator (1,200 pounds of carbon dioxide annually), plug in a hot tub (4,000 pounds a year), or purchase a plasma television, that is their decision. Despite the price increases of recent years, electricity remains a relative bargain. Nonetheless, if stronger building codes, smart grid investments, and conservation programs could reduce forecast demand growth from 2 percent to 1 percent per year, it would be much easier and cheaper to achieve any emissions-reduction goal.

Building New Generation

Time is a critical resource. If citizens want to dramatically reduce emissions by 2050, it's important to get started. But on such a long journey, with so much at stake, it's equally important to get the pace right, too. For Colorado to reduce its emissions by 20 percent, utilities will need to install many billion dollars of low- and carbon-free power plants. In one possible scenario, where future load growth averages 2 percent, the electricity now produced by approximately 2,700 megawatts of coal plants would need to be replaced with generation from five new natural-gas plants, costing $2.5 billion or more. In addition, the state would need to site, finance, and construct large wind farms, perhaps costing more than $6 billion, and build two of the world's largest concentrating solar plants, at a cost of $4 billion. As an alternative, in lieu of the wind and solar, the state could build two large nuclear stations costing roughly $12 billion. These numbers are rough estimates, but they suggest the gargantuan scale of the investment required. Developing a politically viable, fiscally responsible, and logistically practical strategy for making these large investments is an essential first step, but such discussions have hardly begun.

The Energy Landscape of the Future

Since the energy dilemma—in all its dimensions—will not be solved overnight, it helps to take the long view. According to scientists, a 20 percent reduction in greenhouse-gas emissions is not sufficient to resolve the climate challenge. Deeper cuts are needed. Both Governor Ritter and California governor Arnold Schwarzenegger have proposed 80 percent reductions by 2050. This would require enormous alterations in the energy landscape. What would a decarbonized world look like, how do we get there, and what role would electricity play in it?

The Electric Power Research Institute has studied these questions and reached some intriguing conclusions. To achieve an 80 percent reduction in global greenhouse-gas emissions, the burning of gasoline, diesel, and natural gas in a billion vehicles and buildings will gradually need to end, because there is no feasible way to capture emissions from so many distributed pollution sources. In the future, a great deal of fossil fuel might still be used—but only at power plants where carbon-capture technologies are available. By 2050, electricity's share of primary energy will have soared. Carbon-free electricity would need to be everywhere, much like oxygen. It will be produced by some combination of hydropower, nuclear, solar, wind, tidal, biomass, and geothermal, plus coal and natural-gas plants that sequester their emissions.

If the climate challenge is successfully addressed, by 2050 the electric sector will be thoroughly decarbonized, with a carbon intensity that is one-tenth that of today's most efficient coal plants and one-fourth that of today's best natural-gas plants. Because they would not have boilers or furnaces, buildings would be superinsulated. Some would harvest electricity with solar cells on their roofs. Heating and air-conditioning would come via ground-source heat pumps powered by clean electricity. In the transportation sector, electric vehicles would have replaced those equipped with internal combustion engines. Air travel and long-distance road transport would likely use advanced biofuels.

Many technological and economic uncertainties exist. Will carbon sequestration prove economical? Will nuclear power enjoy a renaissance? If sequestration and nuclear power remain economically uncompetitive, perhaps due to a breakthrough in solar technology, then in 2050 the world might be running primarily on flows of wind, water, and sun rather than on fossil fuels. High-voltage direct-current lines might transport solar electricity from the Southwest and wind power from the Great Plains to both coasts.

Reliance on intermittent wind and solar power implies expanded electrical storage in batteries, in flywheels, or as hydrogen. Excess power from wind farms might be used to compress air in underground caverns for later recovery, a technology being explored in Iowa. In mountainous states, the need to store days' worth of electricity might require numerous pumped hydro systems. Smart grid technologies, distributed generation, and super-efficient appliances would be widespread.

If this is where we need to go, policies that focus narrowly on emission reductions tend to obscure what's needed. What we are really talking about is rebuilding the energy basis of our civilization, at a cost of trillions of dollars. If concerns about national security, prosperity, and the environment prompt us to undertake that monumental task, then the investment may well be worth it. After all, there are very large costs in *not* acting on climate change too.

Holy Cross Energy's calculations suggest that meeting the governor's 2020 goals might cost a typical Colorado household an extra $25 to $35 per month, real money to be sure, but something less than an average cell-phone bill. Commercial and industrial customers would see proportionately larger hikes. Are Coloradans willing to pay this much to protect the climate? Perhaps someone should ask them.

A Sensible Strategy

If Colorado aims to decarbonize its economy over the next four decades, then the state government should develop a comprehensive plan to that end. A detailed blueprint and realistic cost estimates are needed, recognizing that fixed targets and timetables may not be the most cost-effective pathway. Building on work by the Colorado Climate Action Panel, Colorado Energy Forum, and Rocky Mountain Climate Organization, a bipartisan stakeholders' group could identify strategies for overcoming the obvious (and not so obvious) economic, technical, and legal challenges.

Simple arithmetic shows that a continued reliance on conventional coal is incompatible with climate protection. However, most of Colorado's existing coal plants will reach retirement age between 2020 and 2040 and will need to be replaced regardless. To avoid stranding valuable assets, the oldest and least efficient plants should be retired first. Public Service Company of Colorado's decision to close its Cameo and Arapahoe plants signal that this process is under way.

By 2050, most of Colorado's electricity will need to come from non-emitting technologies. There are many options, but little certainty about what these replacement resources will cost, where they should be located, who should build them, and when. Operational issues must be addressed. Today's coal plants typically operate 24/7 and will need to be replaced with generation assets that have roughly similar characteristics. Clearly, some new combined-cycle natural-gas plants will be necessary. It's likely that another 2,000 megawatts of wind power will need to be added in eastern Colorado. New transmission lines will be needed to facilitate that, and to tap the solar potential in the San Luis Valley. As time goes on, distributed generation may play a growing role, particularly if there's a breakthrough in solar photovoltaics.

Although it would be difficult to complete the job in a decade, it is possible to make Colorado's energy systems cleaner and more sustainable. Climate change is a particularly taxing challenge that is compounded by the growth dynamic. In Holy Cross's view, resolving it will most likely occur as part of a long-term bipartisan effort to secure America's energy and economic future.

See **James R. Udall**'s biography on page 179. Holy Cross Energy's white paper on reducing Colorado's electric-sector emissions can be found at www.holycross.com.

GETTING THE FEAR
By Todd Neff

As I type these words, electrical pulses in a notebook computer somehow translate the mechanical thrusts of my fingertips into New Times Roman on a flat-screen monitor. The laser printer hums to my right, the dishwasher and washing machine rumble from different corners of the house. There are lamps, phone chargers, a garage refrigerator/freezer keeping unhealthy foods and bags of elderberries in a state of suspended animation. The furnace just kicked on, turning Rocky Mountain methane—four hydrogens mobbing a carbon, derived from photosynthesizers that haven't bagged a photon in 50 million years—into blue flames and parched, sustained gusts from grated holes in the floor.

Fifty gallons of water imprisoned in a cylindrical cauldron endure a scorching with every attempt to equilibrate with the cool 64°F of the basement utility room, such that, at a few moments' notice, oatmeal might be rinsed posthaste from a pink plastic bowl.

When I accompany my daughters to their Montessori school, I bring along 5,600 pounds of fossil-fired Chrysler Town & Country. When I return, I employ a device that uses a tiny propeller to atomize beans grown thousands of miles south of Denver, and another to heat water somewhere below the boiling point to strip vital stimulant from the carnage. And I think nothing of it, generally, any more than I spend time thinking about all those liters of blood flowing through my body, or about what my pancreas is up to at the moment, or that I'm breathing.

Energy is so fundamental, so abundant, so pervasive, it has no real meaning to us. Noticeable only when it disappears, it is classic infrastructure. But we have no sense of energy as a thing in itself, like a road—only its products: light, heat, work, motion.

I am breathing. And after a few years of having covered climate change for the Boulder *Daily Camera*, the effects of our collective energy, if rarely energy itself, do occasionally enter my mind. I know that when I inhale, my lungs take in 20 percent more carbon dioxide than when I took my first breath in 1968.[1] That's 20 percent more of the atmosphere's dominant greenhouse gas today than half a life span ago.

Without carbon dioxide, our planet would average a chilly 3°F rather than our current balmy 59°F.[2] With much more of the carbon dioxide than we have today—38 percent higher than preindustrial levels and climbing—Earth changes its constitution in ways we can't completely know but which, it's widely agreed, will be unfortunate for civilization. Skeptics point out that the planet has been through a roller coaster of carbon dioxide fluctuations—and hence temperature ranges—in its ancient past. They're right. But the changes happened over millennia, giving biological systems time to adapt. We're thickening our greenhouse blanket by roughly half a percent a year, a rocket shot in geological terms. Our sustaining systems, particularly industrial agriculture, are finely tuned to our current global thermostat setting. We fiddle with it at our peril.

Perhaps because my feet are a mile up, I worry less about sea-level rise than about glacier disappearance in the Himalayas and the Andes, which could draw the Yangtze and Brahmaputra and Rímac down to summer trickles, parching the world's most populous places and motivating desperate acts with global consequences. But water supply is just one of an overwhelming litany of climate-change threats (heat waves, droughts, desertification, ocean acidification, more intense storm systems, sea-level rise, extinctions...) upon which experts in the climate-science mecca of Boulder and elsewhere continue to elaborate. Staying abreast of the climate issue is an exercise in being alternatively sobered and terrified. We are almost certainly in worse shape than we think we are, and global carbon emissions are rising faster than the worst-case estimates of even just a few years ago.[3]

We have to do something about it. This is now a consensus among reasonable people. Doing something about it has mostly to do with energy. The Rocky Mountain West is central to any discussion of energy in America: we're home to Powder River basin and other mines supplying 54 percent of the nation's coal, wells producing about a quarter of our natural gas, some of the best solar and wind resources in the country, and the National Renewable Energy Laboratory (NREL) in Golden, the world's foremost green-energy research center.[4] We are a big part of the problem, and will be a big part of solutions to our fossil-fuel fix.

Despite a lot of impressive studies and good intentions (the Priuses, the hemp grocery sacks, the compact-fluorescent lightbulbs), we as western-ers remain as ignorant as America at large to the monumental scale of the challenges we face in transitioning, in any meaningful way, to a true New Energy Economy, as Colorado governor Bill Ritter calls it. By *meaningful*, I mean cutting carbon emissions at a ratio equivalent to those necessary to stop dangerous climate change.

Given a system as complex as the Earth's, nobody can be sure what that threshold is. Climate scientists such as National Aeronautics and Space Administration's (NASA) James Hansen have chosen a 2°C (3.6°F) average global warming. Scientists figure atmospheric carbon dioxide concentra-tions of more than about 450 parts per million risk such heating. That's just 17 percent above where they are today.

Earth's carbon cycle does ultimately flush airborne carbon dioxide and sequester it through geologic processes, but carbon dioxide tends to hang in the air for centuries. Keeping the number below 450 parts per million means cutting carbon dioxide emissions 60 percent to 80 percent by midcentury.

What would reducing my carbon output 80 percent mean? I think about scarcely roughed-up coffee beans producing lukewarm swill, two-hour workdays, a January thermostat set to 42°F. About driving my preschool-ers 25 percent of the way to school—to the convenience store at Eleventh Avenue and Yosemite, roughly, from which they'd walk, through rain, snow, and sleet, like little postal workers.

Consider the Energy Information Administration's (EIA) US forecasts through 2030, which, although conservative, reasonably represent the continuation of American life as we know it without the societal upheaval (which thin, lukewarm coffee would certainly precipitate) or some unantici-pated technological or political development (a $30-per-ton carbon tax, say).

Carbon emissions will climb, the EIA predicts, a total of 7 percent by 2030, a modest amount considering anticipated gross domestic product growth of 76 percent by 2030.[5] Renewable energy looks promising, the EIA says, possibly providing 13 percent of our energy by 2030, more than a doubling of today's sum. The agency predicts solar power exploding from a roughly 1 gigawatt total capacity in 2007 (the capacity, when the sun shines brightly, of a single nuclear plant) to 13 gigawatts in 2030, and wind power from about 16 gigawatts to 44 gigawatts. Assuming these intermit-tent sources generate electricity 30 percent of the time, that's a savings of the equivalent of perhaps fifteen coal plants the size of Xcel's under-construction

Comanche 3 plant outside of Pueblo.[6] Alas, the EIA expects the equivalent of another fifty-three Comanche 3–sized coal plants to be built in this country by 2030.[7]

The EIA is telling us, then, that anything remotely resembling status quo of energy use and delivery stands a good chance of triggering dangerous climate change, unless the rest of the world completely shuts off the lights. So let's look elsewhere—to the Alliance for Climate Protection's "Repower America" program. This is Al Gore's movement to generate 100 percent of America's electricity through renewable sources in ten years. Wind would provide 27 percent of all electrons (1.3 percent today), solar about 16 percent (0.1 percent today; the bulk of the envisioned additions would be in concentrating solar power, which uses mirrors to boil liquids to generate steam to turn turbines), geothermal energy 3 percent (0.2 percent today)…and "negawatts"—energy efficiency—28 percent. There's more, but we can stop.

"It's not feasible in any possible way I can see," said Paul Komor, faculty director of the University of Colorado Energy Initiative and as sharp an energy expert as I've come across.

Wind is ready for prime time, Komor says, roughly cost-competitive with new coal or natural-gas plants. But the very places wind developers covet—those with driving, sustained winds—tend to be sparsely populated because they have driving, sustained winds. So you need transmission lines—perhaps 12,000 miles of them—which cost about $1.5 million per mile.[8] That adds 15 percent to wind's cost, Komor says. Transmission lines require land, and land decisions tend to be made locally—town boards, county commissions, city councils. So wind's challenges are also political and bureaucratic. Still, Komor and NREL experts view the idea of 20 percent or more of US electricity coming from wind as entirely feasible—but probably in twenty years rather than ten.[9]

As for solar energy, solar-photovoltaic power from silicon panels costs about $0.25 per kilowatt-hour, or roughly three times as much as wind energy. Solar energy only makes strict economic sense when massively subsidized, at least without carbon taxes, which beautify renewables by throwing (perhaps deserved) mud at fossil energy. Concentrating solar power is less expensive, costing just twice as much as wind, roughly, Komor says, but also requires transmission from large solar farms in brutally sunny places. Storing superhot liquids to stoke power generation after the sun goes down reduces variability of solar power output, but it adds further to cost.

But let's assume we get to 27 percent wind power and 16 percent solar power, and stretch the deadline to 2030. Forty-three percent of our electricity would come from the sun and the wind. Why not? In 1988, who could have imagined a technology as magically distracting as the iPhone? But even if every renewable kilowatt-hour supplanted a pound of coal (and it won't—the first to go will be natural-gas plants, which are marginally more expensive and emit 40 percent less carbon per unit energy), we'd only wipe out about a third of US carbon emissions. We'd be less than halfway home.

I could go on. Many, many groups have done studies discussing the potential of various forms of renewable energy.[10] I'm familiar with just one that took a serious shot at putting it all together. Chuck Kutcher, a group manager in NREL's Center for Buildings and Thermal Systems, was chairing the American Solar Energy Society's Solar Conference in 2006 in Denver. As part of the program, he asked colleagues from various realms of energy research to come up with their best estimates of how far their various technologies might reduce carbon emissions by 2030—concentrating solar power, photovoltaics, wind power, biomass, biofuels, geothermal energy, plus building efficiency and plug-in hybrid vehicles. It ended up in a book, printed with soy-based ink on recycled paper, called *Tackling Climate Change in the US*.[11] In the foreword, NASA scientist Hansen wrote: "If we are to keep global temperatures from exceeding the warmest periods in the past million years—so we can avoid creating 'a different planet'—we will need to keep atmospheric CO_2 to a level of about 450 parts per million."

Striking in this paper is the role of energy efficiency—fully 57 percent of carbon reductions would come from squeezing more light, heat, and work from less energy. The report discusses improving the efficiency of heating, cooling, lighting, and appliances; using passive solar heat and natural daylight; making more efficient building shells and lighter vehicles and aircraft; and employing more efficient industrial motors, heat recovery, and cogeneration in industrial settings.[12]

Sure, but 57 percent? I look around my house and see squiggly lightbulbs everywhere already. I could unplug the garage fridge, air-dry clothes, set the thermostat to "uncomfortable," install a condensing furnace and a tankless hot-water heater. The real problem is the building envelope—the house is thin-walled—two-by-fours rather than the four-by-sixes better builders use (and Denver code now mandates). I can't send my house over to the gym to bulk up, and stapling Styrofoam to the exterior would get me in trouble with the homeowner's association. Lowering my personal carbon footprint 80 percent would mean serious sacrifice. Every keystroke contributes to the problem. Yet I type on.

When Franklin D. Roosevelt issued "the only thing we have to fear..." sound bite in his first inaugural address in March 1933, hunger ran rampant amid 25 percent unemployment rates. Adolf Hitler had just celebrated his appointment as German chancellor by incinerating the Reichstag. The postwar global order was collapsing. A paralyzing fear shrouded the country.

Now, we have no fear. Not because we are courageous, but because we are complacent. Rather than fear, we worry, a form of fear too faint to motivate, as counterproductive as catatonic terror.

I wrote a book about a space mission. A young aerospace engineer mentioned to me that, at some point late in the mission, everybody just got the fear. The fear that if they didn't work insane hours and sacrifice their personal lives, their spacecraft would stumble and a $300 million effort would implode in their hands. Manhattan Project scientists and engineers had the fear too—that Germany would develop The Bomb first. Apollo's creators, beyond the fear of technical failure haunting all engineers, feared the Soviets would beat them to the moon. The legendary Iowa wrestler Dan Gable trained harder than anyone else because he had the fear his opponents were working even harder. Show me greatness, and I'll show you fear.

What does this have to do with climate and energy? We as westerners, as Americans, must achieve greatness in the coming years, leading a mass transition to energy systems based on lower- rather than higher-density inputs (civilization has always climbed the energy-density ladder, from wood to coal to oil and gas to nuclear), a feat unprecedented in human history. We will have to adjust in many ways, from practicing conservation to the point it feels like rationing to paying far higher prices for everything involving energy, which is everything. A new energy infrastructure will cost trillions of dollars.[13] Adding to our burden is generational injustice: we must atone not only for our own energy irresponsibility, but for the unwitting combustive transgressions of centuries of forebears burning lumps of coal, lighting kerosene lamps, driving big stupid cars with tailfins. We have every reason to fear the consequences of our profligacy, of our continued disregard for the planet's carrying capacity.

Yet there is no mustached villain to galvanize our collective will to act— just a colorless, odorless gas and a lot of dour trends and dire predictions.

A nation distracted by professional sports and popular culture when not working (or, increasingly, job hunting) has little chance of getting the fear. Absent fear, there will be no mobilization and no meaningful action to stem the risk of dangerous climate change.

"If we're really going to reduce carbon emissions by 80 percent, we're going to have to get new technologies online, because it's going to be a lot easier than getting so many people to change their behavior," Lisa Dilling, a professor at the University of Colorado's Center for Science and Technology Research, told me. "People aren't going to voluntarily decide to use less energy and change their lifestyles. It's going to be more a kicking and screaming type of thing."

Without deep carbon cuts by the masses—directly or through taxes or fees to bankroll decarbonization—we are betting civilization on the emergence of magical new renewable-energy technologies.

Perhaps it's a good wager. Ray Kurzweil, the inventor and futurist (among other innovations, Kurzweil invented the synthesizer, thereby enabling futuristic music), believes technology is advancing according to a Law of Accelerating Returns, sort of a Moore's Law for Everything.[14] In the late 1980s, Kurzweil foretold the explosive growth of the Internet in the 1990s and that, by 1998, a computer would defeat the human chess champion. IBM's Deep Blue defeated Garry Kasparov in 1997.[15]

In 2008, Kurzweil predicted that, within five years, breakthroughs in nanotechnology would render electricity from solar panels as cheap as fossil energy, and that in twenty years all our energy will be carbon neutral. I still have the fear, but I now harbor just a hint of blind hope.

Todd Neff is a Denver-based writer. He got the fear while he was science and environment reporter at the Boulder *Daily Camera*. His website is www.toddneff.com.

Notes

1. Available at ftp://ftp.cmdl.noaa.gov/ccg/co2/trends/co2_annmean_mlo.txt.
2. Jeffrey Bennett, Megan Donahue, Nicholas Schneider, and Mark Voit, *The Cosmic Perspective*, Fourth Edition (San Francisco: Addison-Wesley, 2007), 295.
3. Michael R. Raupach, Gregg Marland, Philippe Ciais, Corinne Le Quéré, Josep G. Canadell, Gernot Klepper, and Christopher B. Field, "Global and Regional Drivers of Accelerating CO_2 Emissions," *Proceedings of the National Academy of Sciences* 104 (24): 10, 288–93, www.pnas .org/content/104/24/10288.full.pdf+html. The current baseline for such information, spanning some 2,700 pages, is the Intergovernmental Panel on Climate Change's 2007 Fourth Assessment Report, available at www.ipcc.ch. The *Daily Climate* (www.dailyclimate.org) provides an excellent daily e-mail summary of the world's top climate-related stories. The Pew

Center on Global Climate Change (www.pewclimate.org) and the Union of Concerned Scientists (www.ucsusa.org/global_warming) are also good references. The National Oceanic and Atmospheric Administration's Global Monitoring Division has the latest on global carbon dioxide levels (http://www.esrl.noaa.gov/gmd/ccgg).

4. Available at www.eia.doe.gov/oiaf/aeo/pdf/appa.pdf, p. 30 (coal); http://tonto.eia.doe.gov/dnav/ng/ng_prod_sum_dcu_NUS_m.htm (natural gas, including New Mexico, Colorado, Wyoming, and Utah).

5. Available at www.eia.doe.gov/oiaf/aeo/pdf/appa.pdf, pp. 36, 39.

6. Ibid., p. 33; Comanche 3 will be 750 megawatts.

7. Ibid., p. 20.

8. Available at www1.eere.energy.gov/windandhydro/pdfs/41869.pdf, p. 114.

9. Available at www1.eere.energy.gov/windandhydro/pdfs/41869.pdf. Note that "just" 20 percent wind would amount to an additional 305 gigawatts both onshore and offshore. It would involve adding 16 gigawatts of new capacity (10,700 standard 1.5-megawatt turbines, roughly equal to the total installed US wind capacity today) each year after 2018. Wind farms would cover 19,300 land acres, more than five times the area of Yellowstone National Park.

10. The American Council on Renewable Energy's "The Outlook on Renewable Energy" (January 2007) discusses many such efforts: www.acore.org/files/RECAP/docs/OutlookonRenewableEnergy2007.pdf.

11. Available at http://ases.org/images/stories/file/ASES/climate_change.pdf.

12. Available at http://ases.org/images/stories/file/ASES/climate_change.pdf, p. 47.

13. Britain's Stern Report estimated the costs of dealing with the disasters of unmitigated climate change to be from 5 percent to 20 percent of global annual gross domestic product each year; the Intergovernmental Panel on Climate Change's 2007 Fourth Annual Report estimated the cost at 5 percent annually by 2050. The Stern Report estimated the cost of maintaining carbon dioxide levels below 450 parts per million to be 1 percent of gross domestic product to 2100, or about $500 billion a year in today's dollars—roughly the size of the US defense budget. See Roger A. Pielke Jr., "An Idealized Assessment of the Economics of Air Capture of Carbon Dioxide in Mitigation Policy," *Environmental Science Policy* 12 (2009): 216–225 (available at http://sciencepolicy.colorado.edu/admin/publication_files/resource-2716-2009.03.pdf).

14. Moore's Law: the theory that semiconductor density (and thus processing power) doubles every eighteen months, which Intel cofounder Gordon Moore proposed in 1965 and which has held true.

15. Available at www.nytimes.com/2008/06/03/science/03tier.html?n=Top/Reference/Times%20Topics/People/T/Tierney,%20John.

PIONEERING SUSTAINABLE BUSINESS
By Catherine Greener

A deep spirit of pioneering shaped Colorado. This is about being the first—the first to introduce a new process, product, or way of doing business. This spirit is still maintained by Colorado businesses. Colorado companies and their leadership are pioneering not across vast plains but across new and emerging business models: creating green jobs and green markets. Through four companies in four very different industries, their businesses are examined through a lens of sustainability. Each one of these companies is demonstrating exemplary sustainable behavior along at least one of four dimensions: product, process, governance, and infrastructure. Additionally, each of these companies is working within the principles of natural capitalism and is providing leadership and demonstrating what can be done. Incremental changes will not provide the innovations required for us to significantly alter the course of climate change we are still on. It will take bold, urgent, and courageous leadership to stabilize and protect our resources for generations (and Coloradans) to come.

These four Colorado companies may have little in common at first glance, but they are forging through, making their businesses more sustainable with each transaction. They are working examples of a framework of business practice that was initially developed by the Rocky Mountain Institute and Paul Hawken, and most recently updated by Hunter Lovins and Natural Capitalism Solutions. Principles of natural capitalism are to (1) buy time through radical resource efficiency, (2) reinvent everything—use nature as a mentor, and (3) restore.* A truly sustainable company by this definition would be one that does not leave any waste and provides restoration. Many companies are aggressively reducing their environmental footprints and greenhouse gas emissions and improving their social performance, but a truly sustainable company has yet to emerge. Pangea Organics,

* Adapted from the text on the Natural Capitalism Solutions website, www.natcapsolutions.org.

New Belgium Brewing Company, Namasté Solar Electric, and ProLogis are sustainability pioneers, leading the way and demonstrating possibilities and hope, and often the first to implement sustainable practices.

Product

Pangea Organics

Pangea Organics, located in Boulder, Colorado, is one of the fastest growing companies providing sustainable personal care, or, as they call it, eco-centric skin care. The company strives to have the highest standards for sustainable manufacturing, distribution, and marketing, and may be the best example of a company using all the principles of natural capitalism. Their products are spearheading the sustainable personal care category.

Pangea Organics was founded by Joshua Onysko in 2001 with a mission to "make things that make things better." When asked, why soap? he responds that soap is a product that we need (why make something we don't really need?) and that all the ingredients can be grown. The company's products are manufactured according to the rules of always and never: always nurturing, handcrafted, organic, fair trade (whenever possible), and cruelty free, and never using petroleum, sulfates or detergents, synthetic preservatives, artificial colors or fragrances, or genetically modified organisms.

These stipulations extend beyond the company's products and are used as an education and awareness mechanism to educate a mass market and, by working with other personal care product manufacturers, even change how products are made.

The hypothesis behind Pangea's all-natural ingredients is that there are multiple benefits from a single use—for example, aromatic soap cleans, moisturizes, and potentially improves your mood and senses. The cold processes used to manufacture the soap maintain the integrity of the ingredients that would be lost if the process included high temperatures and boiling. Pangea's niche is the company's claim that their products are "whole organic herbal extracts and whole organic essential oils." Many companies have some botanicals and natural ingredients, but due to the lack of a national organic cosmetic and skin care standard, significant variation of claims to the shoppers exist. Pangea is hoping to lead the industry and demonstrate 100 percent sustainable products.

Pangea's product responsibility extends to the packaging as well. Governed by a zero-waste design parameter, the company pioneered a 100 percent biodegradable, compostable, and plantable package and has chosen glass for several products because of the material's recycled and recyclable closed-loop potential. Consideration is also given to adhesives for the packaging and labeling—many of the packages are designed to be folded and not use any adhesives.

Pangea Organics is changing the landscape of personal care products. If the company is successful, they will fulfill their vision of making things better and provide consumers with opportunities to learn about sustainable practices, the life cycle of products, and the impacts of their choices.

Process

NEW BELGIUM BREWING COMPANY

Not all companies can begin with the founders hiking in Rocky Mountain National Park with a jug of beer in one hand and pen and pad in the other to capture the vision and the core values of what is now one of the fastest growing regional craft breweries in the country, New Belgium Brewing Company. Not only do they have great beer, they have a mission that supports their deep commitment to the environment.

The innovations and continuous improvement at NBB are a direct result of what is called their sustainability management system, which focuses on carbon-footprint reduction, water stewardship, closing loops, and advocacy.

When New Belgium conducted an energy audit of their process back in 1998, they discovered that at the time, the company's largest contributor of greenhouse gas emissions was the electricity that was needed to make the beer. The story, as told in their sustainability report, is as follows: "As a result [of the energy audit], New Belgium employee-owners voted to dip into their bonus pool to subscribe to the City of Fort Collins' wind program at a premium of 2.5 cents more per kWh than fossil-fuel electricity [or 57 percent more at the time]. Thus New Belgium Brewing became the country's first brewery to purchase 100 percent of its electricity from wind power in 1999."

New Belgium then went on to assist the City of Fort Collins in erecting wind turbines to supply the company with additional wind power and allowed for Fort Collins Utility to become Colorado's first electric-utility provider to offer wind power.

Closed loops and industrial symbiosis are the underpinnings of just some of the process pioneering that New Belgium has led. A by-product of New Belgium's (innovative biodigester) wastewater-treatment process is methane gas, the largest component of natural gas. Methane gas is a greenhouse gas that is up to twenty-four times more damaging than the better-known carbon dioxide. The methane gas is captured and returned to the facility, where it runs a process that supplies the plant with both heat and power. The sludge waste from the wastewater-treatment facility is then sold for composting.

Most companies think waste occurs primarily in the solid form, but not at New Belgium. Their focus on closed loops and the elimination of waste also includes an innovative process to recapture waste heat from the brewing process to preheat incoming water that is used in the brewing process. Capturing and using this waste heat eliminates a greenhouse gas–emitting process of preheating water through a process such as a boiler. These types of examples of radical resource efficiency and eliminating waste throughout the manufacturing process can be implemented in most manufacturing processes, even if the company is not making award-winning beer.

Governance

Namasté Solar Electric

The governance of most companies today evolved from influences stemming from maritime commerce, feudal land laws, and stockholder rights from the nineteenth-century railroad expansion. It's been a great 500-year (or so) run. Many claim that the governance of companies has led us to our current state of unsustainability. If a company is going to be truly sustainable, then it needs to go beyond the what and the where it does business—it must address the who. Who is the company responsible for, who does it answer to, and how will all benefit? The infrastructure of the company must support and encourage individual and collective innovation and support the stakeholders and surrounding community.

Namasté Solar Electric was founded in 2005 by Blake Jones, Wes Kennedy, and Ray Tuomey and is the leading installer of solar photovoltaic panels in Colorado. Without the assistance of outside investors, the company grew and maintained their commitment to values while retaining full control of their vision.

Namasté Solar's mission is "We work in Colorado to propagate the responsible use of solar energy, pioneer conscientious business practices, and create holistic wealth for our community."

Namasté Solar is employee-owned. As described on their website, this idea of all for one and one for all means that "as co-owners, we share in the risk, responsibilities, and rewards of business ownership. Co-ownership includes personal responsibility as well as empowerment and encouragement of everyone on the team. Co-ownership is a willingness to continuously hold a bigger-picture vision amidst the everyday details of our individual job roles as part of our continuous improvement."

Namasté Solar staff members have the option to pay into the company to join the ownership team, and the company offers low-interest short-term loans to make this possible. Important decisions at the company are made as a team. Transparency is at the center—when all the employees are owners, there are open discussions about profits and losses, growth plans, investments, and strategies.

The co-owners of the company have different titles to reflect their different responsibilities, but all receive the same compensation and participate in a group-consensus decision-making process. When employees have a voice in the direction of the company, some very attractive work-life practices emerge. For example, Namasté Solar provides six weeks of paid vacation per year for all its employees.

Is this a sustainable model? That is, can a company over a long period of time grow and succeed with this type of unconventional governance?

There are hundreds of emerging solar companies across the United States (over 200 in Colorado alone), yet it was Namasté Solar that was singled out by the Barack Obama administration as an example not only for the 2009 stimulus package, but for how to create green jobs.

All too often, the individuals dedicated to protecting the environment and reversing climate change focus on products and process. Namasté Solar is a pioneer in achieving sustainability sustainably.

Infrastructure
ProLogis
Headquartered in Denver, Colorado, with over 1,500 employees, ProLogis is one of the largest developers/owners/operators/managers of distribution facilities worldwide. ProLogis recognizes that buildings are a significant contributor to climate change. In 2006, ProLogis was one of the first in the industry to adopt a series of sustainability goals for their property portfolio with an aggressive target for 2010. The following are their goals:

- Utilization of 20 percent recycled content, based on cost, in all new warehouse developments
- Diversion of 75 percent of construction debris from disposal in landfills and incinerators on all new projects
- Installation of renewable-energy sources that have a combined generation capacity of over 25 million kilowatt-hours per year across the company's global property portfolio
- Reduction of potable-water usage for landscape irrigation by 50 percent in accordance with methodology established under the US Green Building Council's (USGBC) Leadership in Energy and Environmental Design (LEED) program

To continue their commitment to reduce energy and resource use while reducing climate-change impacts, ProLogis committed in 2008 to two additional sustainability initiatives. They pledged to develop all new warehouses in the United States to standards developed by the USGBC and to register the buildings under the LEED standard. For their international operations, they have also committed to develop new warehouses in the United Kingdom according to the Building Research Establishment Environmental Assessment Method.

As the world's largest owner of distribution centers and facilities, ProLogis has over 500 million square feet of hot, flat roofs worldwide. To reduce waste and increase efficiency, designers and engineers ask the question, who wants what you don't want? ProLogis reframed the question for their business and asked, "Who wants hot, flat roofs?" The response was "How about utility companies in urban centers who are looking to install solar power?"

Thus, ProLogis leases roof space to utility companies, who then fund and install the solar system. ProLogis then oversees the system and the utility pays a monthly rental fee to ProLogis for the roof space. In 2008, ProLogis announced the first of these partnerships with Southern California Edison (SCE). SCE leased over 600,000 square feet of roof space and installed enough electricity generation to power more than 1,400 homes for one year. The second project, announced in October 2008, in partnership with Portland General Electric, is one of the largest solar projects in the Pacific Northwest. Solar photovoltaic cells will be placed on three ProLogis facilities in Portland, Oregon, potentially generating 1.1 megawatts of power. At the end of 2008, ProLogis had over 6 megawatts of projects under development or installed worldwide.

Sustainable Pioneering in Colorado

A tremendous amount of work has been done by the business community to contribute to the field of sustainable business. There are numerous examples of companies using less energy, less raw material, and less water. Some would argue that if we had another 500 years or so to continue this trajectory of incremental and continuous improvement, doing less bad, we would be able to reinvent commerce without much disruption. We don't have that luxury. Our climate and our world are changing rapidly around us. The spirit of sustainable pioneering is what is needed now. Individuals and companies who are pursuing the bold experiments are proving that success of the company and success of the environment are not inverse relationships.

The first principle of natural capitalism emphasizes the need for radical resource efficiency to buy time. But the companies highlighted here are going beyond the first principle and are finding ways to reinvent their products, processes, company governances, and infrastructures. These organizations are exploring virtually the unknown and claiming these new territories as viable ventures.

Despite their differences, all four of the companies have similarities. They have been guided by strong visions, goals, and commitments to creating a sustainable future. They have opened up new areas of thought and possibility, making appropriate adaptations based upon the lessons learned through their sustainable experiments, and have openly shared their adventures with others. They have a strong sense of "what if?" and inspire others in their industry to follow the paths they have created.

Sustainable pioneers such as Pangea Organics, New Belgium Brewing Company, Namasté Solar Electric, and ProLogis are providing the leadership and paths for other companies to quickly follow and find their own pioneering spirit to reinvent everything and restore our climate and planet for generations to come.

Catherine Greener is chief executive officer and founder of Greener Solutions, Inc. Her previous positions included vice president of Sustainability Consulting at Saatchi & Saatchi S, team leader of the Commercial and Industrial team of the Rocky Mountain Institute, and director of quality and customer focus for a division of ABB. Greener lives in Boulder, Colorado, in a house featured on the Boulder solar-home tour.

GREEN SPARK—
ADVENTURES OF A GREEN VENTURE CAPITALIST
By Martha Records

In November of 2007, I hung out my shingle as a Cleantech investor and created a small venture capital firm called Green Spark Ventures. Since that time, dozens of engaging entrepreneurs have walked through our door. Hundreds of people—scientists, salespeople, professors, electrical engineers, lighting designers, architects—have told us their stories of how they hope to fight climate change or battle the continued degradation of our natural world through the spread of their technology. Learning about the market problem they are trying to solve, the technology they plan to develop, hearing their business proposition, their hopes, their dreams, has been wonderful. A privilege, really. I have been struck by the abundance, the true bubbling up of Cleantech innovations.

The Cleantech space is fairly unwieldy. Spanning the entire spectrum of possible end users and including technologies in areas as far-ranging as renewable-energy generation, energy-efficiency infrastructure, water, energy storage, transportation, building materials, plastics, chemicals, recycling, and waste, the Cleantech sector is hard to contain. Since opening our doors, we have met with such a variety of companies. Two Colorado start-ups have developed new energy-efficient window technologies. Another focuses on the conversion of organic wastes such as cow manure into pipeline gas. An Idaho Springs company, Oberon, is working on converting liquid waste from breweries and food processors into protein meal that could be used as a replacement for fish meal in aquaculture. In Boulder, EEtrex converts hybrid vehicles into plug-in hybrid-electric vehicles and is developing a vehicle-to-grid bidirectional charger. Porous Power, a company based in Lafayette, has developed a laminable battery separator that can reduce production costs, enable faster charge times, improve battery performance, and increase the cell cycle life of lithium-ion batteries. And there are so many more.

Green Spark has met with thirty-five Colorado companies and a handful of companies on the West Coast. We have heard from dozens more at industry conferences and Cleantech gatherings. This is just our experience in Colorado, a small slice of the larger Cleantech market. I doubt there is any precedent to this full-throttled enthusiasm on the part of entrepreneurs for creating such a wide range of companies with the dual purpose of making a profit and contributing to the social good. At times it feels almost like a war effort, but instead of it being led from above by charismatic figures, it is advanced by the individual efforts of idealists and businesspeople who are frequently one and the same.

The extent of the entrepreneurial activity in this space is particularly impressive given the difficulties these emerging companies face. They must try to secure capital. They must find and hire scientists willing to work for below-market pay until operating funds are secured. They must assemble a competent management team, protect intellectual property, create ground-breaking innovation, and, in many cases, do so without the certainty that any real market will ever exist for their product. Although Cleantech entre-preneurs operate in widely disparate business fields, what most of them share is a desire to make a difference. And that is where we find common ground.

I stepped into the world of Cleantech investing because I want to play a role in battling climate change. I know that I am a late arrival to the cause. Scientists, teachers, children, parents, businesspeople, and citizens of many countries sounded the alarm on this issue many years ago. Yet it wasn't until after former vice president Al Gore started traveling around the country with his very large PowerPoint presentation that I really paid attention. To be precise, Gore didn't bring me to the table. It was my daughter, at the tender age of ten, who showed me the way. She is a great lover of nature, of mountains and oceans and all the creatures that reside there, and it was her concern about climate change that persuaded me to stop passively observing the efforts of others and get to work on this problem.

Like all parents, I want a bright future for my children. I love the wild places in my adopted state of Colorado and in so many other natural areas I have enjoyed, and I want them to endure. I am concerned about the national security of my country and the resiliency of our economy. And I am worried about the future of those raising their children in poverty around the globe. For all these reasons, I believe that the United States must step off the side-lines and enact meaningful climate legislation. By this I mean we must enact laws that will make oil, gas, and coal more expensive and therefore more painful to use and accelerate the shift in our economy toward conservation, energy efficiency, and renewable energy. I eagerly await the passage of such legislation. I'm not holding my breath, but I'm hopeful.

In the meantime, what to do? As individuals we can make the personal choices that will contribute to carbon-emissions reduction. We can compost, carpool, buy energy-efficient appliances, better insulate our homes. We can ride our bike to work, or if that seems too difficult or dangerous, drive a hybrid or take the bus. We can eat less meat—or no meat—and power our homes with renewable energy. My family does many, but not all, of these things with the goal of reducing our carbon footprint every year. But the cumulative effect of individuals voluntarily making small choices like this will not be enough to get us to our destination. If our goal is truly to have an 80 percent reduction in carbon emissions by 2050, we must have not only a shift in personal behavior, but also a great advance in energy efficiency and renewable-energy technologies to effect significant change.

We will need dramatically more efficient cars, markedly more efficient buildings, restructured utilities with demand-side technologies, a significant reduction in the cost of renewables, and more. All of these gains require the development of new technologies, and this takes time and money. If governments price the externalities associated with the use of fossil fuels, the world's financiers might stampede into investment in these technologies. But as we wait for more government action, I firmly believe that private-sector investing can help get the ball rolling. Driven by idealism or government incentives, consumers are demanding these technologies. And businesses looking to cut costs are interested as well. Many potential Cleantech technologies can thrive even without the promise of future regulatory or legislative changes. If we can increase investment in the development of Cleantech technologies now, then we will be that much further along when our elected officials manage to get their job done.

While my passion may be outsized, the size of my fund is not. I funded Green Spark myself, and it is therefore much smaller than the large Cleantech venture capital funds on both coasts. We have a staff of two: my partner, Dave Ryan, and me. And we have a large, diverse, and very enthusiastic population of possible companies to evaluate. In order to make the most of limited funds and manpower and seemingly limitless investment opportunities, we have had to focus our efforts. Over the last year and a half, we have learned a great deal and refined our approach in the process. Here are our rules.

Play to Your Strengths

I am not a scientist. I can't remember how I satisfied the pesky science distribution requirement when I was in college, but I know it wasn't by taking an actual core science class. And my physics lab notebook from high school is considered by the members of my family to be one of the most hilarious mementos of my youth. My business partner did take a science class in college, but he isn't a scientist either. It is unlikely that we would sniff out the unique opportunity provided by a new development in materials science. For Green Spark, investing in a seed-stage science experiment is not a great strategy.

My background is in consulting to small businesses and start-up companies and private investing outside of the domain of technology. Our strengths are an ability to evaluate the market opportunity, dive into the financials, and assess the strength of the management team.

Stay Local

I know there are rich investment opportunities in California, Texas, and Massachusetts. But local investments are easier to both vet and monitor and therefore more likely to succeed. And because of the presence of the National Renewable Energy Laboratory and the Colorado School of Mines in Golden, the strong engineering departments at the University of Colorado in Boulder and Colorado State University in Fort Collins, and the abundant supply of experienced entrepreneurs throughout the Front Range, Colorado is bursting with Cleantech start-ups. Because venture firms on the coasts seem to have the same preference for nearby portfolio companies, Colorado's emerging companies receive less attention from out-of-state venture firms than they probably should. This makes the need for Colorado-based Cleantech venture firms such as Green Spark even greater.

Recognize the Limitations and the Blessings of Our Fund Size

At its outset, a start-up company will often rely on friends and family and a few angel investors for the needed investment dollars. After this seed stage of development, a company will usually cast a broader net to find early-stage investors who will provide funds needed for activities such as bringing the product to market. This intermediate stage of funding is often difficult for early-stage companies to find. They must turn to potential investors other than friends and family but are often too small to capture the interest of larger institutional venture firms. Green Spark is small enough that we can specialize in this early stage of institutional investing. And there are certain advantages to investing early. By doing so, we can have a greater impact on the development of a portfolio company. And through continued follow-on funding, we can continue to place bets on the company, or not, depending on how well the company performs.

Invest in Companies That Can Succeed
and Prosper under Current Market Conditions

There are a number of Cleantech start-ups developing technologies that are ingenious, elegant, and terrific for the environment but at the same time will only be adopted in the marketplace once some shift in the energy universe occurs. Perhaps their product will be popular with homeowners when utilities charge a low price for off-peak electricity and a much, much higher price for peak-demand electricity. Or perhaps the company's technology will be broadly adopted once electric cars are commonplace. Unfortunately, neither of these market conditions exists currently. While I applaud their vision of a better future, Green Spark does not have the resources to carry them along until the day the market will support their idea. We must search for companies that can prosper now.

Look Closely at Energy Efficiency

A few years ago at a Rocky Mountain Institute (RMI) gathering, an RMI scientist told my daughter, "You have to eat your energy-efficiency vegetables before you can have your renewable-energy dessert." It sounded pretty hokey at the time, but he was right. A recent RMI study found efficiency alone could cut 30 percent of US electricity use and avoid the need for 60 percent of coal-fired generation. In the struggle to reduce our nation's carbon footprint, the potential of energy efficiency is enormous. By reducing our total demand for electricity at the same time that we build up renewable-power generation, we can eat away at the carbon-emissions problem from both ends.

In addition to being an important part of the climate-change solution, energy-efficiency technology companies are often those Cleantech companies that are best positioned to succeed in the near term. While the purchase of a rooftop solar array may pay for itself in a period of eight years (or more, depending on where you live), the purchase of energy-efficient lighting may pay for itself in less than two years. And in today's economic climate, energy-efficiency technologies targeted toward the commercial and industrial markets do particularly well as businesses search for ways to increase profits by cutting costs.

In February of 2008, Green Spark made its first investment in an energy-efficiency technology company called Albeo Technologies. Based in Boulder, Albeo is a solid-state lighting company that designs and manufactures white-LED lighting fixtures for the commercial and industrial market. Even in the midst of the current economic downturn, Albeo's sales are growing. They have a terrific management team, a track record of customer-focused product innovation, and a product line that can save their customers thousands of dollars in the long run through reduced energy and operating costs. Albeo's LED lighting fixtures have no bulbs that need changing, last more than fifteen years, and use less energy than comparable incandescent or fluorescent fixtures. Some customers install Albeo light fixtures in public spaces to demonstrate their sustainable practices, but many install their fixtures in seldom-seen data centers or warehouses in order to cut energy and bulb-replacement costs. Our investment in Albeo feels like a great beginning.

As we look forward, I hope for two things. I want the companies we bet on to succeed financially and further the important goal of reducing carbon emissions. To hit both of those targets would be truly gratifying. But I also hope that one of the Colorado companies pinning their hopes on a highly experimental idea, a company that we didn't have the capacity to support, will go on to be a game changer. Their success in contributing to the fight against climate change will make it easier to laugh about the big one that got away.

Martha Records is a Cleantech investor and the founder of Green Spark Ventures. She lives in Denver, Colorado, and enjoys exploring the Rocky Mountain region with her husband and three children.

FIRST, DO NO HARM—
GREEN BUILDING AND THE PRECAUTIONARY PRINCIPLE
By Josh Radoff

Somewhere in New York City in early 2008, a still-solvent major multinational company finds itself in the early throes of adopting a sustainability platform. They had just taken the dramatic step (at least perceived by them) of mandating Leadership in Energy and Environmental Design (LEED) certification for every new building that they planned to develop across North America and, if the effort went well, eventually rolling out a similar program internationally. My firm, which consults on sustainability issues for clients like this, was assisting in the rollout, providing a variety of technical and strategic support along the way, and on this occasion I found myself among upper management presenting an educational session on green building and on the mechanics of the LEED program for such a large portfolio of projects. When the formal presentation was over and we began the question-and-answer session, one of the corporate executives raised his hand and asked a very direct question. "This is all well and good," he said, "but when does all this end? When do we get to stop being green?"

A good question…

What he meant was that, if they went ahead and successfully implemented the program—an effort that would force their company and all of their development partners (the architects, the engineers, the contractors, the subs) to change the way they built buildings, not in a fundamental or profound way, but in a way that would significantly improve the performance of the buildings, the health of the occupants and visitors, and the impact on the environment—it would take a major effort on the part of this company and add cost to the way they did business, at least in the short term. (In the long term, the hoped-for goal was to make green building standard practice and return the cost premium to zero.) And it wouldn't happen overnight. Achieving the goal would likely take at least a year or two of hard work with much pain and frustration with the various constituents (this much has turned out to be true). What this executive was doing, therefore, was looking ahead to that point in the future when they might choose to uncork some champagne and celebrate in environmental ecstasy their successes and look back with satisfied grins on the hundreds of new facilities that they had built to LEED standards and that have the coveted LEED-certification plaque adorning the front doors. "What then?" he was thinking... Would they consider their job well done and then proceed in this new fashion into the foreseeable future? Would they be heralded as a sustainable company? Or would they already be sadly behind the green wave and in the position of having to resume the process all over again, this time pushing even deeper? And if the latter, when would it end?

That question, and more importantly the answer, have been at the forefront of the minds of not only those in the sustainability arena, but also everyone else who either cares about the state of the world and their role therein or perceives the risk of being branded an irresponsible laggard or a great green pioneer. Where are we going, how do we get there, and where does it end? Is this the movement du jour, or is it the new paradigm that will continue to evolve and re-create itself for the next 100 years, at the least?

The answer to the last question seems the most clear. If it were a flash in the pan, the trend would have peaked and begun to descend by now, and the recession would have only hastened its decline (as it did in the 1980s for the efficiency and solar movements). Instead, there is no end in sight to the work ahead.

The question then becomes what and how. Five years ago, the forefront of the green-building wave centered on development projects pursuing, and hopefully achieving, LEED certification. Now it's the pack—the mainstream of the real estate and development industries—that is catching up: building projects of all shapes, sizes, locations, and budgets are pursuing LEED; cities and states are mandating LEED for all municipal projects; and three major US cities are now requiring LEED certifiability[1] for private-sector projects as well. But rather than marking the end point of the conversation, the mainstream's entry into the greenery has only served to open some eyes and start the wheels of a deeper discourse turning.

Much of LEED, and in fact the vast ocean of environmentalism that we've known to date, have been a steady progression toward the concept "First, do no harm." We got the picture that our patterns of development and methods of building were doing all manner of harm, and so we decided to try to do the logical thing: do less of it. Use less energy, use less water, consume fewer resources, produce less pollution—such that if we follow the linear progression of this goal, we simply end up at the point where doctors pledge to start: "First, do no harm." This mantra is encapsulated in the highest standards of green building and environmentalism: zero-energy buildings, carbon-neutral buildings, LEED-platinum buildings. Given where we've come from, if we can achieve these goals, then we will have come a very, very long way. But even a zero-energy building may or may not achieve the aim of doing no harm. If the building is located in the suburbs or accessible only by car, then by making the building zero energy, it has first created a serious energy and climate problem and then made efforts to solve one small piece of it.[2] What's more, the fundamental direction of doing less is purely reactionary. It focuses on what we don't want rather than what we do want.

So what do we want?

There are some who see the future as defined by the dominance of a new technology, the way urban planners saw the future defined by the automobile in the 1950s. Wind farms, fuel cells, solar panels, algae-derived biofuels, smart grids—these have all captured the imagination of the environmental community, and some view the present challenge as simply developing these technologies to rescue us from ourselves like some mechanical messiah. The ideas aren't that far-fetched either. We know, for example, that the amount of solar energy falling on a south-facing roof of a house in Denver is enough to fuel the entire house. The problem is that the solar panels that we'd use to collect the energy are very expensive relative to the cost of electricity and natural gas. So if only we could convert more of that incident solar energy to electricity (photovoltaic panels are generally around 10 to 14 percent efficient, meaning that 86 to 90 percent of the energy is not captured) or change the technology so that it simply cost less, we'd be in great shape. Or if we could only expand our transmission system to accommodate vast amounts of renewable energy from massive centralized solar-power plants in Arizona and Southern California and find some way to store the energy not produced on cloudy days, we could have a carbon-free electrical grid with no nuclear power. Imagine that.

For others in the green-building and sustainability world, the next wave isn't about technology, although technology is certainly a central tenet. The reason for this is that they see the myriad issues we're facing as a society not as independent, but as fundamentally linked. How does a solar panel, for example, impact how your food is produced or reduce the 2,000 to 3,000 miles it traveled to get to you? How does it impact the way we devote our landscape to automobile traffic (and how can that land be used otherwise)? How does it lead to species diversity? How does it alleviate the issues of social or environmental injustice? And how does it impact human health and well-being?

Surprisingly, a solar panel can impact all of these things to one degree or another. If every US household installed enough solar-collection capacity to power itself, it would create massive jobs and help to add beneficial productivity to the economy. It would reduce our need to extract and produce coal and natural gas, thereby alleviating the pressure on and pollution of the land and waterways where they come from. It would mitigate the impacts of climate change and the associated displacement of poor people, the loss of agricultural land, the spread of disease, and the loss of species around the world. In other words, it would do a lot of good.

But there is also a trend toward more explicitly recognizing the interlinking nature of both our problems and our priorities and goals. If, for example, we recognize that we are healthier and happier if we have access to natural landscape and daylight; if we eat fresh, nutrient-rich nonprocessed foods; and if we exercise, walk to get around, have access to quiet places, get out of our cars, then we start to see our building and development projects very differently. Development, after all, is literally the "bringing about of potential or possibilities." With this in mind, it makes us ask the question, what is the potential of this piece of land, this urban infill site, this building, this community? And then the follow-up question, why are we doing this and what do we hope to achieve?

Some in the green-building industry have started to ask this question about their projects. Their clients tell them they want an office building. They ask why. Because they want a place to house their workforce. Why? Because they want a place where everyone can come together and be productive in getting their work done. Because they want to develop a community where people are aligned with the goals of the company and are inclined to stay there. Because they want people to be healthy and feel good about coming to work. Because they know if their people are happy, then they will have a good chance of being successful in their endeavors in the long term, all of which would make them happier too.

So what they need is not an office building, but a place that can create community and wellness, foster creativity and productivity, and bring out the best potential of their people and their company. This changes quite a bit how we think about the project. It's no longer a building. It's a vehicle for something else entirely.

The challenge, of course, is knowing what to do once you've asked and answered these questions. Maybe it means incorporating an area on your site or roof for agriculture, or shifting the placement and orientation of the building to allow for natural ventilation, passive heating and cooling, and daylight. Maybe it means restoring the natural hydrology of the site—allowing wetlands to do the job of mechanical and chemical storm-water and sewer treatment, or bringing to the surface a river that had long ago been buried. Or if you are a city planner, maybe it means creating neighborhoods and cities that are built to human scale and provide for green space, community gardens, farmers' markets, and developing relationships with regional food growers for community-supported agriculture programs. Maybe it means allowing for additional density, devoting less land to parking, providing for affordable housing, creating habitat and wildlife-migration pathways, and providing enough mixed uses like schools and grocery stores so that one can live with only the car available through the community car-share program. Maybe it means considering your storm water to be your drinking water (which it is) and treating it as such, and considering the need to plug into a fossil-fuel-driven electrical grid an unconscionable act.

The themes present in all of this tend to blend the urban, agricultural, and natural landscape, using the lessons of biomimicry and permaculture, such that there is no place where the city ends and nature begins. Instead, the urban landscape softens, buildings play the role of natural landscape, and neighborhoods become a mix of buildings, small farms, vegetation, and transit. The Living Building Challenge[3]—a concept gaining in traction and appeal—begins to embrace these concepts, at least at the level of the individual building. It effectively asks that a building act like a tree: that it use only the energy incident on the site; that it use only the water that falls on the site and that the water leaving the site as clean as when it arrived; and that the materials be harvested and reused somewhere else or nourish the ground they decay in.

Like so many things, this trend is essentially re-coupling the elements that we have historically decoupled. As Michael Pollan points out, by taking the animals off of our industrial farms, we created two discrete problems: a need for fertilizer, which we solve by using nonrenewable fossil fuels, and pollution in the form of runoff from feedlots, which we haven't addressed at all.[4] By re-coupling these issues, we turn waste into food and start closing the loop. Similarly, by re-coupling the urban and natural landscape, we create the potential for human well-being and reengage with the natural landscape and the soil and, of course, continue to do no harm.

So it's not just about technology or the will to put it into place and spend the money on it—though admittedly, without a better solar panel, building automation, and window, we wouldn't get very far. It's about envisioning how development can unlock the potential of the place and about working toward what we actually want, rather than backing away from what we don't.

Josh Radoff is cofounder and principal of YRG Sustainability, based in Boulder, Colorado. He is a regular speaker on sustainability issues and has consulted on hundreds of sustainability projects at the intersection of the energy, climate, and green-building fields, both nationally and internationally.

Notes

1. Not actual LEED certification, but rather an awkward and at times confusing LEED equivalency.
2. For the average American office building, the energy consumed by the building occupants traveling to and from the building is significantly greater than the energy required to run the building itself.
3. Developed by the Cascadia Chapter of the US Green Building Council (www.cascadiagbc.org/lbc).
4. *New York Times Magazine*, October 2008.

THE CHANGING RULES OF ENERGY FINANCE
By Matthew H. Brown

The typical activist government or business response to evidence of climate change is to commit to reduce carbon emissions. Many go further and make specific commitments to buy power from renewable-energy sources or to reduce their energy consumption. A few take other steps like buying carbon offsets to effectively neutralize their emissions of greenhouse gases. These are critical steps. But an essential element is missing—one that is a necessary step if the fight to stabilize carbon emissions is to be successful: financing. This chapter describes a few innovations in financing the transformation to a low-carbon economy. These innovations do not rely solely on government funding; such funding will always be limited and will never be sufficient to stimulate the massive changes that will be required for this transformation. Instead, these innovations focus on exploring ways to leverage private and public capital, using both funding sources together so as to multiply their combined effects.

The historical background upon which these innovations rests began in the late 1970s in the United States when states like California and the federal government adopted policies to encourage renewable energy. Climate change was not yet on the policy agenda at the time; energy security was paramount. Some of these early policies worked successfully over several years to nudge emerging renewable markets closer to the mainstream. California's combination of tax credits and standardized utility payments to independent renewable-power generators in the 1980s is an example of this kind of financing policy. The financial returns from these California policies alongside federal tax incentives meant that developers of renewable-energy facilities could often make money from the incentives alone, whether or not their facilities actually produced any energy. Fundamentally, though, such a policy is not sustainable; it relies on a business model that is entirely dependent on subsidies, and subsidies themselves rely on government's unpredictable appetite to consistently extend those subsidies.

Ideally, the financing arrangements to support energy efficiency and renewable energy should rely less on subsidies and more on sustainable financial partnerships between governments and the private sector to provide low-cost capital for clean energy. To be really successful and sustainable, however, any public-private financing initiatives should follow four rules. The initiatives must be:

- Scalable. They must be able to step up to a very large scale; pilot programs are fine, but the goal should be to mobilize large amounts of capital and deploy that capital through well-marketed programs that reach large numbers of people.
- Secure. Initiatives must recognize that money put at greater risk will cost more than money put at lower risk. Therefore, financing programs should be structured to provide secure ways to recoup money and to distribute risks to those who can best bear those risks.
- Sustainable. Initiatives that rely on rebates may have a short-term impact on building a market for technologies that are more expensive than their traditional competitors. However, a goal for financing should be to seek sustainable sources of funding that can provide steady support to the investments that the private sector and governments must make over a period of many years.
- Simple. Simplicity is key. People must be able to borrow money through a streamlined process that performs quickly and minimizes paperwork while adhering to underwriting standards necessary to ensure that borrowers will be able to repay their loans. Financing programs, whether designed for homeowners, developers, industry, or government agencies, should be designed for simplicity.

Following are three brief profiles of new public-private partnership financing mechanisms that support renewable energy or energy efficiency. Although they are still best described as pilot-scale efforts taking place in a just a few communities or states, they are useful examples.

Example One: Mortgage Financing for Homebuyers and Refinancing

Mortgages are long-term financing agreements secured by the home that they finance. Despite the turmoil in the housing markets that became so evident in late 2008, mortgage financing has a lot of potential to benefit energy efficiency and renewable energy. A thirty-year mortgage paid over 360 monthly payments provides for a long amortization period, so that the incremental monthly cost of solar panels or energy-efficiency upgrades is small, and those costs generally become tax-deductible if included in a mortgage.

The idea of financing energy efficiency and renewable energy through a mortgage is not new. Starting in the early 1990s, Fannie Mae and other federally established mortgage agencies created a new energy-efficient mortgage product through which the agency agreed to buy mortgages that take into account the fact that the ratio of borrowers' income to their expenses improves when their utility bills decrease. An energy audit and certifications were required to qualify for this mortgage. The effect of this better ratio has been to enable homeowners to qualify for a slightly larger mortgage than would otherwise be possible. In part because of the paperwork involved in qualifying for these mortgages, and in part because the benefit of qualifying for the mortgage was relatively small, few people took advantage of these mortgages—the federal government insured only a few hundred such mortgages nationwide in 2007.

A small cadre of states is now experimenting with a new type of energy-efficient mortgage product designed to be more flexible and useful than the old product. Created by the US Environmental Protection Agency (EPA) in coordination with the nonprofit Energy Programs Consortium, this product does not rely on any reassessment of projected utility bills or recalculation of income-to-expense ratios. Instead, it requires only that the home either meet an ENERGY STAR® standard or that an existing home's energy efficiency improve by at least 20 percent, in the case of an energy-efficiency retrofit. The EPA is now operating a pilot program in which it gives lenders certification as an ENERGY STAR® lender if their mortgage product is based on these standards and provides consumers with a verifiable financial benefit such as an interest-rate reduction.

Colorado is one among several states that offer this program. The state's energy office has partnered with the Bank of Colorado to buy down one discount point on a mortgage for a new ENERGY STAR®–qualifying home, and the one-point reduction can be used to reduce the mortgage interest rate. The Bank of Colorado and the state match funds to buy this discount point reduction. Through the combination of the rate reduction and projected energy savings, the homeowners can immediately save money—despite the higher cost of the house. Colorado and the bank designed the program to serve the energy-efficient new-home market, but it could be structured to serve the home-refinancing market for energy-efficient upgrades in new homes.

This program has the advantage of leveraging state funds—in this case doubling state funds through the bank's match—and offering the financing in a streamlined way through existing financial institutions.

Example Two: State Treasurer Money Invested in Energy Efficiency

Governments take in money through tax and other revenue sources and spend it for government operations such as salary, utilities, or capital purchases. But governments are also money managers. One responsibility of state treasurers, for example, is to manage their state's funds by placing that money in secure investment pools that will yield a return to the state. These funds can be enormous; California's employee pension fund is worth over approximately $200 billion, for example. These funds also represent an opportunity to provide capital for investments in clean energy. States treasurers in Pennsylvania and Colorado now fund multimillion-dollar energy-efficiency loan programs.

The two programs are similar. In the case of Colorado, the Colorado State Treasurer agreed to capitalize a fund with an initial investment of $4 million per year but has not defined the number of years that the program will run. Pennsylvania created a three-year program valued at slightly more than $6 million per year. That money will be invested in a loan fund to support energy-efficiency loans that are typically in the range of $7,000. A private lender originates and services the loans, then sells them to the treasurer—so the treasurer in essence acts as a secondary market for the loans. The state also provides the lender with access to a loss reserve to guard against potential loan defaults, while the lender responsible for originating these loans guarantees the loans to the treasurer. The effect is that the treasurer funds have a double-barreled security through the loss reserve and through the lender guarantee.

The key elements of this program are the innovative use of funds with different features and the combination of state funds with private-sector lending and marketing expertise:

- Treasurer funds require a return on investment and must be invested conservatively.
- Other state funds must be used to support clean energy but can be placed at greater risk and do not require a return.

- A private program administrator conducts contractor training and recruitment as well as marketing.
- Lender expertise provides the loan-processing and -origination services.

Example Three: Property Tax–Based Financing Programs

One of the newer innovations designed to raise and deploy capital for energy efficiency and renewable energy began as a pilot program in Berkeley, California, and has quickly spread to other cities in California as well as other states, including Colorado. This new program relies on property taxes as a way to repay a loan made to finance energy efficiency and solar energy. The program requires that borrowers agree to join a voluntary special district (much like a water or fire district through which property owners support local services by a charge on their property tax bill). Upon joining the special district and taking out a loan to finance qualifying efficiency or renewable-energy measures, the homeowner allows the local government to place a lien upon the house. The loan-repayment charges then appear on the homeowner's property tax bill and, if the homeowner moves, are transferred to the subsequent homeowner. This system allows for a lengthy amortization period for the loan. The system also provides for a secure payment stream through the property tax—in case of a mortgage loan default, the property taxes are high on the priority list for repayment. This secure payment stream also has the effect of reducing the cost of money used to finance the loans. Financing can come from private capital sources or from public bonding. Boulder, Colorado, may use private activity bonds, which are federally tax-exempt, to fund its program. Berkeley, California, may use private capital investors to fund the program.

The advantage of this program is that it gives homeowners access to funds that they can repay over an extended period, using the equity in their house as collateral. It also provides an easy and streamlined way for homeowners to repay the loan through the property tax.

The drawback is that it, like many financing programs, is restricted to homeowners who have the ability to borrow money. One more financing tool may provide a solution for this last group of people.

Example Four: Bill/Tariff-Based Financing

Utility consumers in Hawaii, Kansas, New Hampshire, and elsewhere can take advantage of a financing mechanism that lets them pay for energy-efficiency or renewable-energy measures on their utility bill. Sometimes referred to as PAYS, or Pay As You Save, this program allows utility customers to borrow money to pay for qualified efficiency or renewable-energy measures. Utilities loan customers the money, but the customers then pay for principal and interest through their utility bill as part of a special tariff that the regulatory utility commission approves. This approach means that consumers do not have to sign a loan agreement—the charge is simply added to their utility bill. If the customer moves away, the next resident takes up the charges until they are completely paid. Much like the other approaches described in this chapter, this approach allows for an extended repayment period based on the life of the measure rather than the life of a loan.

Capital to fund these programs can come from utilities, from bond issuances, or from public benefit funds (small charges that all utility customers pay through their utility bill, the proceeds of which go to support clean-energy investments). If the capital comes from bond issuances, the fact that utility bill payments are typically a secure revenue stream (in some cases, failure to pay results in disconnection) means that the cost of capital to fund the bonds will be lower than it would be for bonds funded through a less secure stream of revenue.

Summary

Each of these tools represents an example of a government entity seeking new and previously untapped sources of capital to finance either energy efficiency or renewable energy. The net effect of these programs should be to give homeowners or homebuyers a new way to finance investments that will help them save both energy and money. This is important for the substantial portion of the population that may not be able to front the cost of a multi-thousand-dollar investment in solar panels, energy efficiency, or other measures. Ultimately, these financing measures will be critically important to the daunting effort to move beyond the public commitments to reduce greenhouse gas emissions and toward widespread measures that will put those commitments into action.

Matthew H. Brown has worked for twenty years in Europe, North America, and Asia on energy issues. Brown has written more than fifty articles and books on renewable energy, energy efficiency, energy regulation, transmission, energy technology, and critical infrastructure protection. He holds a bachelor of arts degree from Brown University and a master's of business administration from New York University.

THE COLORADO CARBON FUND—TAPPING INTO THE VOLUNTARY GREEN MARKET TO REDUCE GREENHOUSE GAS EMISSIONS
By Susan Innis

Dairy farmers in Colorado are struggling these days; the price of milk has tanked and banks that lend to farms are going under. But still, dairy farmers are interested in doing their part to help the environment. A new technology called an anaerobic digester can capture the methane gas generated by cow poop and use that gas to generate heat and power. Not only do dairy farms get a chance to be energy self-sufficient, but by installing these new technologies on the farm, they can help to reduce the amount of methane being released into the atmosphere. Since methane is one of the six gases responsible for global warming, any technologies that can help reduce methane emissions are good news.

The state of Colorado has a new program, the Colorado Carbon Fund, designed to help support new, local, innovative energy projects like anaerobic digesters. The fund will pay to purchase the carbon offsets from projects that capture methane or use energy efficiency or renewable energy to reduce greenhouse gas emissions. This revenue stream can be used by the project owner to help pay back loans and generally cover the cost of installing a technology. The carbon offsets are quite valuable, as they represent the emissions reductions and environmental benefits of taking methane and carbon dioxide out of circulation. In turn, the Colorado Carbon Fund provides these high-quality carbon offsets to individuals, businesses, and events as a way to help them reduce their carbon footprints.

Many individuals and business owners are increasingly concerned about energy and climate-change issues. However, most folks don't really know where to start or what to do to make a difference, but they are willing to help do something. Even better if they can lower their energy bills and help the environment at the same time. For these folks, donating to the Colorado Carbon Fund to support new clean-energy projects in their home state is a great step toward feeling like they've been able to do their part and contribute to the solution.

How Does the Colorado Carbon Fund Work?

The Governor's Energy Office launched the Colorado Carbon Fund in August 2008 with two main goals. The first was to add some credibility to the world of carbon offsets by providing high-quality offsets from projects located here in the state of Colorado. We think we can help build consumer confidence in offsets as a way to reduce carbon emissions by supporting local greenhouse gas mitigation projects with strong environmental benefits and verifiable emissions reductions. The second goal of the program was to establish a funding source for new, innovative energy efficiency and renewable-energy projects.

The Colorado Carbon Fund is set up as a partnership between the Colorado Governor's Energy Office and The Climate Trust, a nonprofit with a strong track record in developing credible greenhouse gas mitigation projects. The fund accepts tax-deductible contributions from individuals, businesses, and events and uses them to purchase and retire carbon offsets from eligible clean-energy projects. The fund supports technologies that capture methane, such as anaerobic digesters and solar hot water systems that use sunlight to heat water, displace, or reduce the use of natural gas for water heating in hospitals, jails, and recreation centers. The fund is also investigating innovative transportation projects that reduce the use of gasoline or diesel in fleets of vehicles.

Projects are solicited through a rolling open application. The Climate Trust evaluates prospects based on several criteria, including emissions reductions from reducing or displacing the use of fossil fuels. We have to ensure that the project is actually removing or displacing carbon dioxide, so it is important to make sure we do those calculations appropriately.

We also want to support brand-new projects that for a variety of reasons wouldn't happen but for our investment in them. The Colorado Carbon Fund looks for projects that face a variety of barriers. Financial barriers are obviously one. There are a lot of great technologies out there, but they don't always get installed or built because of financial reasons. There are also some other hurdles. For example, here in the state of Colorado, solar hot water is a great technology. We have 300 days of sunshine a year, but we don't have very many large-scale solar hot water projects. That tells us that this technology is facing some sort of a barrier. It might be that facility managers who own and operate the kinds of buildings where that technology might make sense aren't aware of the technology and its benefits. There might not be enough installers who are qualified to install this equipment. Then, too, there is general human nature and inertia—as human beings we don't always go out and try the newest, greatest thing. We're sometimes a little reluctant to change. Good projects often face a variety of these sorts of barriers.

We're also looking for newer technologies that have a great greenhouse gas reduction benefit but maybe haven't been well commercialized yet. And of course, we look for projects that wouldn't be happening otherwise. Anything that would be required by law or that a company or an organization would be doing anyway is not the kind of project that we want to invest in.

Raising Awareness

One general criticism of carbon offsets is that they don't create a strong driver for consumers to change their behavior or purchasing decisions. If they can just write a check, they may think they have solved the problem. However, climate change is a serious and tough challenge and we're going to need to do everything we can to stop it. For this reason, we created the Project C outreach campaign to help educate individuals and businesses about how we can each do our part to help reduce emissions and collectively make a difference in tackling the climate-change program. Our marketing slogan is "Think, Act, Fund," and we encourage individuals, businesses, and event planners to calculate their emissions, take steps to reduce what they can, and then offset through the Colorado Carbon Fund to compensate for what they can't reduce on their own. By viewing offsets as a last resort and encouraging consumers to pursue efficiency, we'll have a much bigger effect on reducing emissions.

As with all programs in the Governor's Energy Office, the Colorado Carbon Fund works in partnership with a number of local governments and community organizations across the state. Local governments have been very strong supporters of climate action, with many developing community-wide greenhouse gas emissions inventories and implementing community-based climate action plans to do what they can to reduce emissions and raise public awareness of the problem of climate change. The Colorado Carbon Fund supports these community efforts by returning a portion of all donations to local community partners by funding outreach, education, and small-scale energy projects.

History of Voluntary Green Markets in Colorado

Coloradans have long had a strong interest in supporting clean-energy projects in their own communities. For the past decade, most of the state's electric utilities have offered voluntary green-pricing programs, whereby folks can opt to pay a bit more on their power bill to support new wind farms. Those programs were instrumental in helping to jump-start the first few renewable-energy projects in the state. They were successful in part because folks could drive an hour outside the Denver metro area and see some of the country's first large wind farms. Citizens felt a strong and direct connection in helping to make those new projects happen.

In 2004, Colorado voters increased support for renewable energy by passing the first statewide referendum on renewable energy. Amendment 37 required utilities to obtain 10 percent of their energy mix from renewable sources like solar and wind. Due to popular support, the state legislature later expanded that policy, and now the state's largest utilities will get 20 percent of their power from renewable sources by 2020. The voluntary green-pricing programs and progressive state renewable-energy policies have led to the installation of more than 1,000 megawatts of wind energy and thousands of rooftop solar panels.

Over the past few years, some consumers have started to question whether paying more on their utility bill can really help drive renewable-energy installations beyond what would be happening anyway in response to state policy mandates. There is a strong interest among some businesses—Aspen Skiing Company, for example—to drive development and innovation even further. Aspen Skiing Company and others have expressed a strong interest in directly helping to fund brand-new, local projects. They want to see that their investment in renewable energy directly leads to development of new projects and they prefer to see those projects developed as locally as possible. Through the Colorado Carbon Fund, we now have a product that responds to this market demand to link a voluntary purchase with a brand-new, local greenhouse gas mitigation project. While carbon offsets are rather intangible, to the extent we can help make the direct link between a company's donation and a new project, we can build more credibility.

There are a number of carbon-offset programs that are run by nonprofit and for-profit companies. Many of them are focused on developing projects overseas, which is terrific for helping the developing world move ahead with innovative technologies. However, here in Colorado, many folks want to have a more direct role in solving the climate-change problem. Since our reliance on fossil fuels has helped create the problem, it makes sense to try to change the way we do things locally.

Going Forward

The Governor's Energy Office hopes that the Colorado Carbon Fund will build capacity within the state to develop greenhouse gas mitigation projects and monetize the reductions to participate in national and international carbon markets. With President Barack Obama's administration's commitment to making the United States an international leader in tackling climate change, the Colorado Carbon Fund can serve as a model for developing expertise among local businesses and agencies in participating in the international market.

Susan Innis is the Colorado Carbon Fund program manager for the Colorado Governor's Energy Office (GEO). Prior to joining GEO in 2007, Innis spent eight years as an energy policy advisor and green-power marketing director at Western Resource Advocates, a regional conservation law and policy center. She holds a master's degree in public administration from the University of Colorado at Denver, studied energy planning and sustainable development at the University of Oslo, and earned a bachelor of science degree in biology from McGill University in Montreal, Canada.

IT'S A WONDERFUL LIFE—
NATURAL GAS AS A BRIDGE TO THE NEW ENERGY ECONOMY
By Michael L. Beatty

Everyone remembers Frank Capra's Christmas classic *It's a Wonderful Life*, with Jimmy Stewart in the role of George Bailey. George has always dreamed of attending the university, designing bridges and skyscrapers, and seeing the world, but he never has the chance. Instead, he remains in Bedford Falls at the Bailey Savings & Loan, protecting his neighbors from the rapacious Henry Potter, the owner of the bank, who abhors the "sentimental hogwash" of homes for the working poor. After Uncle Billy loses an $8,000 deposit and the bank examiner threatens to close the business and prosecute George for fraud, George contemplates suicide. Only after Clarence Odbody, an Angel Second Class, shows George that without him Bedford Falls would have descended to sex, sin, poverty, and perdition, do George, his family, and neighbors realize that it is a wonderful life.

The movie is a modern morality play that illuminates our nation's character, extolling the American virtues of hard work, self-sacrifice, community, frugality, and faith and rejecting the sins of greed, selfishness, envy, and deceit. Given the theme of the movie, there is great irony in the savings and loan scandals of the past decade as well as the current economic collapse caused by the bursting home-mortgage bubble. Homeownership may be a laudable aspiration, but offering a no-money-down adjustable-rate mortgage on a $1 million house to a family with a combined income of less than $100,000 is a bad bet—no matter how highly we value homeownership. Perhaps the real moral lesson should be that values matter but facts rule.

Green energy policy has now taken on the same moral connotation as homeownership. America has made a moral choice for affordable, abundant, carbon-free energy and doesn't want to hear dissent.[1] Unfortunately, turning the energy debate into a moral issue instead of a policy discussion is destructive and keeps us from exercising prudent judgment. Fossil fuels can represent the values of either George Bailey or Henry Potter. Fossil fuels have enabled America to become the most powerful nation in the world, given us the highest standard of living, enabled our agricultural industry to feed the world, and saved us from destruction at the hands of the Kaiser, Tojo, and Hitler. Coal, oil, and natural gas continue to provide steady, dependable, and affordable energy for our nation's homes, commerce, and industry. Likewise, we can argue that fossil fuels are not only destroying the planet, but also dehumanizing humanity. These "fuels from hell" and their greedy sponsors have bankrupted our nation, increased our trade imbalance, fueled Middle East terrorism, funded the Taliban, exploited labor, created monopolies, and precipitated every ill from the Ludlow Massacre to 9/11.

After more than a century of being cheap, plentiful, and ignored, fossil fuels are either the root cause of every problem or the foundation for every solution. Fossil fuels heat our political rhetoric as well as our homes and fuel partisanship as effectively as automobiles. Nowhere is the fight more acrimonious than Colorado. Our plentiful natural-gas resources and strong environmental ethic have caused a bitter fight with no middle ground. Making energy policy a moral choice has polarized the vocabulary such that reasoned discussion is now impossible.

For the great majority of citizens, the contradictory claims create paralysis, typified by the shopper who drove her car home to get Whole Foods cloth bags because she was unwilling to choose between paper and plastic. We need to have a serious discussion about energy policy and we need to make hard choices, but we also need to start that discussion with a clear understanding of the facts and a willingness to be rational.

Those facts start with three important points. First, global warming is caused by burning fossil fuels. In the words of the Intergovernmental Panel on Climate Change (which shared the 2007 Nobel Peace Prize with Al Gore), it is "very likely" that increased carbon in the atmosphere is caused by human use of fossil fuels and is the cause of global climate change.[2] The objective scientific community has reached a consensus. We need to address carbon emissions, not ignore them.

Second, the exponential growth in global population and increase in global health and living standards means that we will need approximately twice the energy we consume today by the year 2050. The United Nations Population Division estimates that world population will increase over the next forty years by approximately the same amount as the total population of Earth in the year 1950. Earth's population will increase from the current 6.7 billion to 9.2 billion, equaling the total population of 2.5 billion in 1950.[3] Not only will there be many more of us, but far more of us will be driving cars, using computers, and eating better than we are today.

Third, we will not be able to impose burdensome costs on energy without creating a black market in cheaper energy or widespread rejection of the energy policy. America can reduce its own burning of dirty fuels by imposing a carbon tax or a cap-and-trade system on carbon emissions, but we cannot force developing countries to follow our lead or prevent the global movement of capital and manufacturing to the cheapest energy alternatives. While a significant reduction in carbon emissions by America is essential, we cannot ignore costs in making energy choices. Coal can generate electricity for a fraction of the cost of large-scale renewable projects. If coal is an unacceptable fuel, the substitute must be a reasonably priced alternative. Markets work, and our energy policy must accept the reality of the marketplace.

Interestingly, these three points embrace the Democrats' concern with global climate change, the Republicans' concern for free and open markets, and the universal acknowledgment of the global population explosion. Just as energy policy is a bipartisan problem, we need to find a bipartisan solution.

The first step in finding that solution is to acknowledge that any current step will require some form of compromise. The dream of a carbon-free hydrogen economy—for example, extracting hydrogen from seawater with solar energy, then using that hydrogen to power fuel cells in cars and homes and leaving only freshwater as a residue—simply does not exist.

Step two requires that we acknowledge fossil fuels as the only option that is currently both plentiful enough and inexpensive enough—even including the cost of carbon emissions—to fuel the energy needs of the planet. Other options are years away from meeting the extraordinary energy demands of the planet.

Step three requires that we acknowledge that step two is not a viable long-term option. There must be a way to immediately and substantially reduce carbon emissions while giving other technologies the time, money, and research to become viable alternatives. Nate Lewis, a chemist at the California Institute of Technology, has done an exhaustive study of our energy options. Lewis's findings have been echoed by both David Nocera, professor of energy at the Massachusetts Institute of Technology, and the late Richard Smalley, a professor at Rice University and Nobel laureate. His study concluded that the planet currently consumes 13.5 terawatts of energy (a terawatt is 1,000 gigawatts, or 1 trillion watts), almost all of which comes from fossil fuels. Meeting the requirement of an additional 13 terawatts of power by 2050 is virtually impossible if we eliminate fossil fuels. If we dam every river on Earth that could technically produce power, we would gain only 1.5 terawatts. If we were to build new nuclear plants at the rate of one every two days for the next forty years, we would gain only 8 terawatts. If we were to convert all arable land to ethanol production, assuming sufficient water to grow it, we would gain 7 to 10 terawatts—but have nothing left to eat. If we were to build wind turbines on every parcel of land with a wind strength of at least class three (wind blowing at an average of 11.5 miles per hour at 33 feet above the ground), we would gain only another 3 terawatts. Lewis, Nocera, and Smalley all conclude that only solar power has the ultimate capacity to provide sufficient carbon-free energy to fuel the world, but acknowledge it is not yet price competitive.[4]

Those are sobering numbers, but it does not mean that all is lost. We can make meaningful change immediately and still give solar and other technologies the chance to develop in a manner that can transform our current energy portfolio. All we need to do is recognize that all fossil fuels are not created equal. America must embrace our vast resource of natural gas.

Coal is the cheapest and most abundant fossil fuel available on Earth. It is clearly the first choice of developing countries such as China and India for those very reasons. But coal is also the dirtiest of the fossil fuels and the largest source of carbon emissions. Moreover, clean coal is like a healthy cigarette—it doesn't exist. We must reduce the world's use of coal, but renewable-energy alternatives that can deliver neither abundance nor affordability are unacceptable. Indeed, moving to expensive alternatives that cannot provide reliability will undercut their viability for the future. Moving to natural gas is the only rational choice.

Natural gas emits one-half the carbon that coal does to generate the same amount of energy, and one-third less carbon than gasoline. Using natural gas instead of coal for power generation and to back up new, renewable-energy projects as they come online would cut our emissions while still maintaining affordability. Using natural gas for fleet and personal vehicles would have an equally beneficial impact on carbon emissions at a price that is lower than that of gasoline.

Why hasn't natural gas been embraced as a solution? First, there is the widespread belief that there isn't sufficient natural gas. Second, people are concerned about the historical swings in natural-gas prices. Third, the natural-gas industry hasn't wished to emphasize the environmental benefits of natural gas because historically many of those same companies have owned either oil or coal interests as well and have not wished to undercut their own products. None of those issues should be problems now.

Just a few years ago, American engineers made a breakthrough that has turned conventional wisdom on its head. The natural-gas industry has learned to produce "unconventional gas" in huge quantities. This gas, from tight sands and gas shales long known to contain huge quantities of natural gas, has only recently become commercially productive. Horizontal drilling and hydraulic fracturing, among other innovations, have unlocked enormous quantities of domestic natural gas—much of it in Colorado. America is now awash in natural gas.

It is true that there are still volatile price swings, but a stable market would support long-term natural-gas contracts and prevent or curtail those swings. Consistent use of natural gas will support the consistent drilling that is required to maintain a consistent price. Finally, many of the new gas reserves have been discovered by independent companies searching for natural gas only. Many of the leading producers of natural gas are now unfamiliar names to most Americans, but they are dedicated to making America aware that natural gas is the best option for reliable energy that will immediately reduce carbon emissions.

Colorado's natural gas is clean, abundant, affordable, and American. It is a bipartisan solution for America, helping to reduce carbon emissions, generating high-paying domestic jobs, providing American consumers with reasonably priced energy, and reducing our dependence on imported oil. Moreover, in an economy that has recently received the right jab of high oil prices and the left hook of a banking and housing collapse, natural gas does not require the trillions of dollars in new infrastructure that renewable-energy options will require. It is not the final answer, but it is certainly the best answer for today.

If Frank Capra and Jimmy Stewart were to remake their movie, I believe George Bailey would run a gas utility. Bedford Falls would be powered with clean-burning natural gas, everyone would have a good job, utility rates would be low, education would be well funded, and research would be supported for a carbon-free future. Only Henry Potter would be furious because his coal plant would be out of business.

Michael L. Beatty is chairman of Beatty & Wozniak, PC, a thirty-five-attorney law firm headquartered in Denver, Colorado, and dedicated exclusively to the energy industry. A graduate of the University of California at Berkeley and Harvard Law School, Beatty currently has an active legal practice and serves as a director of two publicly traded energy companies. Beatty has also been a law school professor, general counsel of a large multinational energy company, chief of staff to Colorado governor Roy Romer, and chairman of the Colorado Democratic Party.

Notes

1. See editorial by Senator Abel Tapia (D-Pueblo), "Pueblo Bishop Arthur Tafoya supported the oil and gas [rulemaking] because protecting the health of our people and our land is a moral issue," *Pueblo Chieftain*, March 12, 2009.
2. Intergovernmental Panel on Climate Change, Fourth Assessment Report (November 17, 2007).
3. Thomas L. Friedman, *Hot, Flat, and Crowded* (New York: Farrar, Straus and Giroux, 2008), 28.
4. Nate Lewis, "Towering the Planet," keynote speech, California Clean Innovation Conference (May 11, 2007), and Richard Smalley, "Future Global Energy Prosperity: The Terawatt Challenge," *Materials Research Society Bulletin* 30 (June 2005).

OIL SCARCITY AND CLIMATE CHANGE
By Steve Andrews

Anyone reading this book probably lives in an industrial society that has taken oil for granted for decades. Yet, we rely on that oil to power millions of now-critical daily tasks that a century ago still would have been regarded as minor miracles, from travel by car to refrigerated shipments by eighteen-wheelers to transcontinental flight.

During the post–World War II era, we got hooked on truly cheap and plentiful oil. From 1945 to 2005, worldwide oil consumption grew tenfold. But in 2005, the easy growth in annual oil production slowed to a crawl, despite generally high prices. While wildly gyrating prices hogged the headlines, this production slowdown became the first unmistakable signpost that our oil joyride is heading for the off-ramp.

To date, our responses to oil supply and consumption challenges indicate that we're focused on two related consequences: climate change and national security concerns over our growing reliance on imported oil. But when developing policy strategies to combat these two consequences, we must understand the looming oil scarcity issue as well. To date, there is limited evidence that we do.

But isn't there an upside here? If worldwide oil production is slated to start declining soon, won't that be a climate-change plus? Maybe, maybe not. It depends on the choices we make, on how we respond.

Oil Flows

Roughly 36 percent of the world's commercial energy comes from oil. While shares for the other fossil fuels—coal (27 percent) and natural gas (23 percent)—are on the rise, the flow of oil proves tough to replace. And that flow, at plus or minus 84 million barrels a day, is enormous.

How big? If you ever cross the bridge over the Colorado River in the western Colorado town of Glenwood Springs during late July, look down. The river rushing below roughly equals the amount of oil the world is consuming at that moment in time.

In the United States, close to 70 percent of the 19 million barrels we consume daily runs our transportation system. Within that transportation sector, the largest share goes to gasoline, then diesel and jet fuel. Oil consumption in power plants declined from 17 percent in the early 1970s to 2 percent today. This means that cutting down on our oil use revolves tightly around our transportation system, not the power-generation sector.

Oil Scarcity

We're not running out of oil, either in the United States or around the world. But we're running out of options to steadily increase the available supply. In fact, before the 2008 recession, it was getting tough just to maintain oil production at then-current levels. Declining annual production in older fields was catching up to the more publicized gains in new fields coming online. Then, after the last fast growth period (2003 to 2004), production flattened.

With increasing frequency, new countries join the unfortunate club of oil-producing nations in which production has slipped into permanent decline. Among the world's twenty largest oil producers (the Big 20), which produce most (84 percent) of the world's oil, the first to decline was the United States (1970), then Indonesia (1977), the United Kingdom (1999),

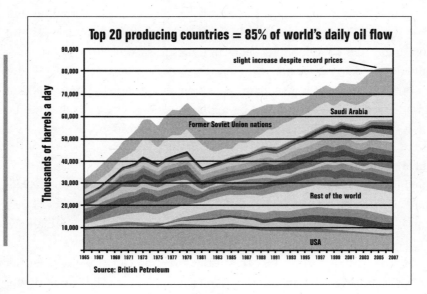

Figure 1

Norway (2001), and Mexico (2004). Figure 1 shows that over half of the Big 20 either are experiencing flat or volatile production (including Russia, Iraq, and Nigeria) or have passed their peak production and are in decline. While the remainder still grow their supply, some nations (including China and Azerbaijan) are approaching their peak production era; declines will follow soon.

Eventually, the math dictates that declines will more than offset gains. At that point, world oil production will have hit an all-time high, a peak. In May 2009, a report by the respected investment analyst firm Raymond James & Associates stated that "peak oil on a worldwide basis seems to have taken place in early 2008." They concluded that "reaching peak oil still represents a transformative moment in the history of the oil market...it is only a matter of time before prices begin to reflect the reality that oil scarcity may become a fact of life in the not-too-distant future."

Raymond James is only the latest to reach this conclusion. In Denver, in the spring of 2009, widely respected oil-industry financial analyst Tom Petrie said, "you can make a good argument" that world oil production won't ever exceed last summer's peak. Capital funds manager T. Boone Pickens agrees. Back in the fall of 2007, Sadad Al-Husseini, retired vice president of Saudi Aramco, the world's largest oil company, stated that world oil production was within a year or two of hitting a final plateau. Christophe de Margerie, chief executive officer of France's oil giant Total, said in the winter of 2009 that world oil production would probably never exceed 90 million barrels per day—a level only marginally above last summer's high (87 million barrels per day). Other companies and organizations identifying oil scarcity as a near-term concern are Toyota, Merrill Lynch, Deutsche Aerospace, Volvo Trucks, the US Army Corps of Engineers, the World Resources Institute, and the nation of Sweden, among others.

Yet, due to the worldwide economic crisis of 2008, caused in part by record-high oil prices, falling demand for oil pushed the looming oil scarcity issue to the back burner. A temporary oil-supply glut during early 2009 whipsawed oil markets into a period of deep uncertainty. Short-term, the producing nations' capacity to supply oil to the world market could exceed demand for several years. Longer-term, the fact that $170 billion of investment in more expensive future supply has been delayed or canceled during a recent six-month period will, best case, cap supply within three or four years. Even the formerly optimistic International Energy Agency warned in August 2009 that investment slowdowns will lead to continued flat production, more oil-price spikes, and a peak in world oil production within ten years.

The Coming Oil Scarcity versus the Optimists

The notion that world oil production is or soon will be rolling over from sufficiency to scarcity draws powerful resistance. Most of the optimists argue that the size of the remaining oil resource is enormous. Their thinking goes like this: in the last 100-plus years, we've burned just over 1 trillion barrels of oil, and we have much more than 1 trillion left in the ground. On top of that, eventually we plan to extract another 10 percent, from old fields, through enhanced oil-recovery methods. Then, too, we have 3 to 5 trillion barrels of unconventional oil—Canadian tar sands, Colorado oil shale, liquefaction of Montana coal, and so on—to extract. So it will be decades before we anticipate any decline in supply.

The lead voices making this optimistic argument are not small fries: the US Energy Information Administration, ExxonMobil, British Petroleum, Daniel Yergin's Cambridge Energy Research Associates, the Organization of Petroleum Exporting Countries, and the International Panel for Climate Change (IPCC). But their ranks are steadily thinning.

From a simple supply-side perspective, the optimists' argument misses the pivotal question. It isn't, how big is the resource? but rather, how fast and how efficiently can industry tap those resources? The answer: the job is getting harder and harder every year, for a whole host of reasons:

- Geopolitical conflicts and resource nationalism restrain access to the Middle East and some of the best remaining conventional oil targets.
- The annual decline rate of existing on-land oil fields is gradually increasing, while the decline rates of large offshore projects drop even faster.
- Production from unconventional oil resources is much harder to scale up than that from conventional oil fields.
- All the unconventional oils cost more to extract, and we consume more energy in the extraction processes, leaving less net energy with which to run society.

At the end of the day, all the work to grow oil supplies from unconventional sources will merely slow down, not reverse, the accelerating decline rates of conventional oil supplies over the next several decades. This is just one of several counts by which the optimists' arguments against looming oil scarcity fall short.

Responses to Scarcity and the Climate Link

There are many possible policy response scenarios. Consider the range of three here. First, if the IPCC is right, there are more hydrocarbon resources available to extract and burn than nearly any other credible organization thinks. In that case, a business-as-usual scenario would see carbon dioxide emissions rising steadily for a century (assuming carbon sequestration for coal-fired power plants never works or isn't implemented). But a growing chorus of critics, including a full panel of presenters at the American Geophysical Union's fall 2008 meeting, challenge the IPCC's wildly optimistic assessments of remaining oil, coal, and natural gas. (See papers by James Hansen, David Rutledge, K. Caldiera, B. Mignone, and R. Brecha.) Don't bet on this one.

Second, and more likely, as conventional oil slips toward scarcity, we may decide to accelerate development of unconventional oil resources. That in turn would ramp up carbon dioxide emissions because it takes much more energy to produce high-quality energy liquids from low-grade oil sands, oil shale, and coal-to-liquids than it does to simply extract an equal amount of conventional oil from the ground. But, as mentioned above, unconventional oil supplies are unlikely to come to market quickly. Barriers include high energy inputs, high water inputs, high financial costs, and a changing political landscape.

Third, if we embark on a crash program to improve transportation-sector efficiency, the demand for oil could flatten to more closely match looming supply scarcity. Then, if we gradually shift parts of the transportation sector to run on electricity—electric trains and plug-in hybrid-electric cars, for example—the push for accelerated power generation by renewable-energy resources stands a chance of slowing and capping carbon dioxide emissions at a lower rate. And perhaps ethanol from non-corn biomass will eventually prove successful at the niche-fuel level. If the large flows of expensive natural gas from shale formations become more cost-effective, we might shift a little more transportation to running on natural gas. But the odds against this best-case scenario, and a lower-carbon path from transportation fuels, are long.

The sad reality is that, in large part due to looming oil scarcity, we're likely to suffer through an enormously bumpy transition. There are flickers of hope for scenarios driven by smarter policy paths than the dumb ones we've chosen in the past (such as massive incentives for ethanol from corn). But with oil scarcity looming, we're running out of another precious resource: time.

What Does This Mean for the Rocky Mountain States?

The Rocky Mountain region, while still a significant oil producer at the national level, is an oil has-been. Among them, New Mexico, Wyoming, Montana, Colorado, and Utah account for roughly 10 percent of US oil production. Yet that supplies only 4 percent of the US daily oil diet. But just like the nation at large, the region's production topped out in roughly 1970; since then, production is down 50 percent (see Figure 2) despite higher prices and application of much better technology today. More drilling in remaining off-limits areas would only slow down, not reverse, regional oil declines. But some drilling and production will continue, regardless of access limits.

From a supply perspective, the search for alternatives to conventional oil focuses on and will generate substantial work in the Rocky Mountain region. Optimistic developers of oil shale's long-standing potential will increasingly probe the Green River Formation, the largest and richest such formation in the world, located in northwest Colorado, southwest Wyoming,

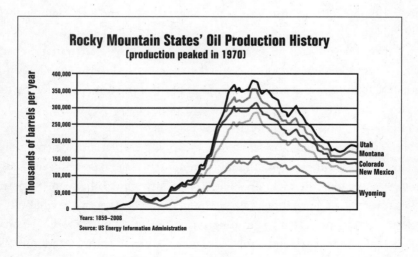

Figure 2

and northeast Utah; that will call for importing more people and delivering more services to one of the country's least populated areas. Proponents of the Pickens Plan call for a large shift to natural gas from gasoline and diesel as a transportation fuel; this would increase regional drilling in one of the most actively drilled portions of the country (from 2000 through late 2008). Montana wants to lead the implementation of coal-to-liquids technology, should it ever become more cost-effective than it is today. And if electrification of transportation becomes a part of our future, development of wind and solar resources in and around the Rockies may ramp up even faster than the current accelerating level of activity.

On the demand side, outside of large urban cores, mass transit is a less compelling investment in the low-density mountain states unless development and redevelopment messages and incentives change dramatically.

Steve Andrews has thirty years of experience in the energy sector in consulting with builders, municipalities, and utilities, as well as working with public television shows and freelance writing. In 2005, he cofounded the nonprofit Association for the Study of Peak Oil and Gas–USA.

PART FOUR

POLICY AND POLITICS

For over thirty years, the Rocky Mountain West has elected state and national leaders who have helped to lay the foundation for the environmental, natural resource, and energy policy innovations of today. Senators Tim Wirth and Gary Hart; western political families including the Udalls, the Salazars, and the Babbits; Colorado governors Richard Lamm and Roy Romer; and many others created a legacy that has been further developed and expanded by today's Senators Tom Udall and Mark Udall, Denver mayor John Hickenlooper, former Salt Lake City mayor Rocky Anderson, Colorado governor Bill Ritter, Montana governor Brian Schweitzer, New Mexico governor Bill Richardson, and others.

The shift in popular interest and concern about climate issues in the past few years means that politicians now run *on* climate issues rather than running *away* from them, to paraphrase a point made in the essay by Jill Hanauer, David Winkler, and the staff of Project New West. They document findings from extensive polling of the electorate in the region across party lines on issues of climate, energy, and environmental policy, with some surprising results.

Chip Ward discusses the security threat we face from environmental crises inside the country and suggests how to tap the resources right in front of us to address them. Florence Williams profiles veterans responding to a similar sense of opportunity (recently enhanced through stimulus incentives). Heidi VanGenderen, the state of Colorado's first climate advisor, compares and contrasts the US approach to public climate messaging and strategy with that which she observed in a six-month fellowship in the United Kingdom. Former *Rocky Mountain News* media observer Jason Salzman explores local attitudes to media reporting on climate issues in a national context. And my closing essay describes why the climate polcies of our cities are more relevant than ever in a national context.

RED, BLUE, AND GREEN—
THE WESTERN POLITICAL REALIGNMENT

By Jill Hanauer, David Winkler, Lisa Grove,
Melissa Chernaik, and Andrew Myers

The Intermountain West—Arizona, Colorado, Idaho, Montana, Nevada, New Mexico, Utah, and Wyoming—has emerged from the relegation of red state status in 2004 to become the new frontier of American politics in 2008 and beyond. Due in some part to the region's rapid population growth, the West has undergone some of the most dramatic political change in the nation, as evinced by Barack Obama's victories in Colorado, New Mexico, and Nevada (none of which were carried by John Kerry); the election of five Democratic governors (a significant improvement over the zero governorships held by Democrats in 2000); a nearly even split on Senate seats—seven Democrats to nine Republicans (Democrats held just three in 2000); and a dramatic reversal in control of congressional representation (Republicans sat in seventeen House chairs after the 2006 election to Democrats' eleven—now the numbers are precisely the opposite).

There are a number of reasons that this political change is occurring in our region, but there is one overarching reason: quality of life. People move to the West and stay here because of the unique quality of life the region offers. Indeed, nine in ten Interior West voters say that they "enjoy a unique quality of life," a characteristic that is deeply valued across all states surveyed. Nearly as many (88 percent) say that their fellow westerners "cherish the outdoors." Interestingly, analytical modeling reveals that the latter underpin the former, in that westerners' perception of a unique quality of life is enhanced by access to the outdoors, which is cherished for the solace and centering it provides. To these Americans, the outdoors represents a combination of physical exercise, recreation, family time without the interruptions of technology, a place of reverence, and a way to get and stay grounded. While climate change is not voters' top issue—that hallowed ground is reserved for more immediate concerns such as job creation, household costs, and education budget cuts—westerners' close connection to the land means that we see climate change as a threat to our way of life, not only from a recreational perspective, but from an economic one. Whether voters are ranchers, farmers, or those whose livelihood is connected to the recreation industry, climate change has a direct impact on their financial security. As a result, candidates now run *on* climate change, rather than running *from* it: the new breed of western elected official is largely willing to confront the issue, rather than deny it, and voters in the Interior West have embraced politicians of both parties who offer pragmatic solutions to climate change.

Yet, as this region has grown, so the Republican machine has grown increasingly out of touch with the values and issue priorities that draw people to the West and keep them here, such as education, healthcare, growth, and conservation, just to name a few. Simultaneously, a new breed of Democrat has emerged in the West, one that is independent, is pragmatic, and governs not with a partisan ideological lens but a western one. This new western Democrat has made it safe for Republicans and unaffiliated voters, the key to the kingdom in many of our western states, to stick their toe in the water and begin voting Democratic up and down the ballot.

The New Energy Economy and Its Impact on Politics and Policy

In the lead-up to the 2008 election, several events—including skyrocketing gas prices, increasing public concern about foreign oil dependence (with its impact on homeland security), and the early rumblings of what would later become a stunning economic meltdown—converged to finally make plain to Americans the connection between the somewhat ethereal issue of climate change and their family's own bottom line. These events resulted in the birth of a new linguistic term, which quickly made its way into the public consciousness and political vernacular. Suddenly, *environmental regulation* was no longer a dirty phrase, and political candidates of all stripes were climbing over each other to be the first to embrace the New Energy Economy, with its promise of lower energy costs and good US jobs.

By late 2007, western voters had begun to draw a clear distinction between environmental regulation specifically and government regulation in general, about which voters were significantly more divided at the time. When asked to choose whether tougher environmental regulations "are worth the costs" or "cost too many jobs and hurt our economy," a clear majority (57 percent) agreed that environmental regulations are worth the price. Nevadans are the most likely to line up in favor of this perspective, shared by two-thirds of the electorate there. Just under six in ten Colorado, New Mexico, and Arizona voters agree with this assessment. While just under half of Montanans share this opinion, they still offer a ten-point margin over the alternative viewpoint that these regulations cost too many jobs and hurt the economy.

Similarly, Nevadans are the most likely to say that the federal government should invest more in alternative energy research. Six in ten Nevadans favor this point of view, and just over half of Arizona, Montana, New Mexico, and Colorado voters agree. Region-wide, 56 percent opt for alternatives

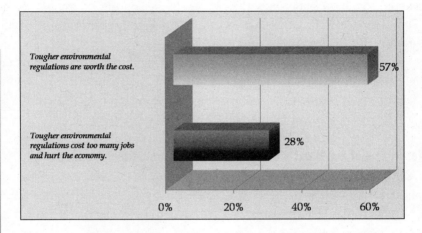

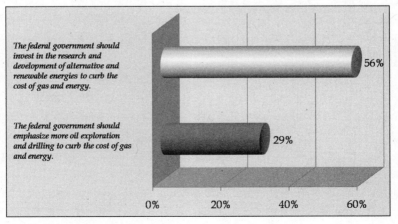

over exploration, while just over one-quarter (27 percent) of voters concur instead with the premise that "the federal government should emphasize more oil exploration and drilling" to curb the cost of gas and energy, giving renewable proponents a nearly thirty-point advantage over the naysayers.

A tangible example of this pro–alternative energy mind-set in action, and its ability to shift the political playing field, can be found in Colorado. After four failed attempts to enact renewable-energy goals in the state legislature, proponents brought the issue to the November 2004 ballot in the form of Amendment 37. Despite millions spent in opposition by Colorado's largest energy utility, Xcel Energy, the amendment passed by a convincing eight-point margin—the first renewable portfolio standard (RPS) law to be passed through the ballot-initiative process in the country. Not only was the effort itself successful, it helped catapult Democrats to control of the state legislature for the first time in forty years.

Further, after meeting the tough standards set in the initiative by 2007 (eight years ahead of schedule), Xcel Energy set aside past differences with Amendment 37 proponents and actively joined an effort to expand Colorado's RPS to 20 percent by the year 2020.

Water: One of the West's Oldest Issues Remains Salient

Over three-quarters (77 percent) of intermountain westerners are of the opinion that residents of their region are "more likely to be affected by water issues." As was true of their affection for the outdoors, water—access to it or lack thereof—helps define the western quality of life.

In addressing these water issues, the western voter strongly prefers regional cooperation to federal intervention. When given a choice, nearly half (49 percent) agree that "solving the dwindling water supply issue in the West is better done by the states in the region, not the federal government," while only one-third (33 percent) believe that "the federal government should do more to solve the issue of dwindling water supplies in the western US."

Senator John McCain learned this lesson the hard way in August of 2008 when he suggested that the 1922 Colorado River Compact (the water-sharing agreement among California, Nevada, Arizona, Colorado, New Mexico, Utah, and Wyoming) "needs to be renegotiated." Reaction in Colorado to the perceived "water grab" was swifter than the river itself, epitomized by then-senator Ken Salazar's comment that "Senator McCain's position on opening up the Colorado River Compact is absolutely wrong and would only happen over my dead body." Republican US Senate candidate Bob Schaffer was equally blunt, saying that renegotiation would happen "over my cold, dead, political carcass."

Realizing the power of the issue, the League of Conservation Voters seized on it, launching an effective campaign centered on a hard-hitting television ad slamming Senator McCain for promoting renegotiation "so Arizona and California can grab our water" and having "a 'disturbing ignorance of the West's scarce water resources.'" The ad closed by imploring, "Let's defeat John McCain and his plan to take our water."

The Denver Post piled on too, editorializing in a "Memo to John McCain" that the presidential contender should "forget about winning our nine electoral votes next November. We don't vote for water rustlers in this state; we tar and feather them!" The Post's predictions turned out to be prophetic: on Election Day, Obama won Colorado by a margin of nine points (54 percent to McCain's 45 percent), not only beating his Republican rival, but far outstripping the performance of the last two Democratic presidential contenders in the state (plus-fourteen points over Kerry in 2004 and plus-seventeen over Al Gore in 2000).

Case Study: Climate Change as an Issue in the Presidential Election in Colorado

Water issues were far from the only conservation concern with an impact on the presidential vote in the West. McCain, a leading Republican advocate for confronting climate change, made a strong effort to lay an early marker on the issue. In a May 2008 campaign ad, McCain told the public, "I believe that climate change is real. It's not just a greenhouse gas issue. It's a national-security issue. We have an obligation to future generations to take action and fix it."

Yet McCain was never able to achieve credibility on climate change in key western states such as Colorado. In early June 2008, Coloradans gave a strong advantage to Obama over McCain when it came to tackling the issues. By a margin of twenty-nine points, voters were more likely to think that when it came to being "committed to action on global warming," Obama (51 percent) would do a better job than McCain (22 percent). Colorado voters also felt that, by a margin of twenty-five points, the Democrat would do a better job "spurring the development of a clean, renewable-energy economy that reduces our dependence on foreign oil" (Obama 51 percent, McCain 26 percent)—an item that drove voter support, according to analytical modeling produced at the time of the survey. And while McCain held a slight advantage (plus-five) on "protecting our unique way of life in the West" (McCain 36 percent, Obama 31 percent), that advantage was likely erased by the Colorado River Compact debacle in August.

Colorado voters' greater trust in Obama to deal with climate change also translated to greater support for the Democrats' plan. By a margin of fourteen points, Coloradans were more likely to agree with the merits of Obama's position on climate change (57 percent) than with a summary of McCain's position (37 percent).

Conclusion: A Critical Issue for a Critical Region

That President Obama's administration chose Denver for the ceremonial signing of the 2009 American Recovery and Reinvestment Act—dubbed by the president as "the most sweeping economic-recovery package in our history"—is extraordinarily telling of the Interior West's continued importance on the national scene. Equally instructive, while the legislation devotes significant resources to popular programs including education, infrastructure improvements, and healthcare modernization, the president and vice president chose to be introduced by the chief executive officer of Namasté Solar Electric, a solar-energy company based in Boulder, at the Museum of Nature and Science during their visit to Denver to sign the economic stimulus bill into law. Further, the president's remarks devoted three full paragraphs

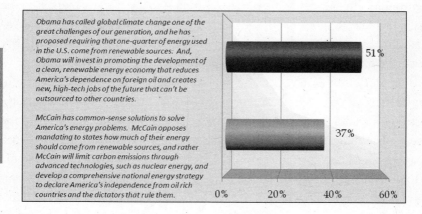

(compared to one paragraph apiece spent on education, infrastructure, and healthcare) to alternative-energy development, boldly declaring:

> Because we know we can't power America's future on energy that's controlled by foreign dictators, we are taking a big step down the road to energy independence and laying the groundwork for a new, green-energy economy that can create countless well-paying jobs. It's an investment that will double the amount of renewable energy produced over the next three years and provide tax credits and loan guarantees to companies like Namasté Solar, a company that will be expanding instead of laying people off, as a result of the plan I am signing.
>
> In the process, we will transform the way we use energy...the investment we are making today will create a newer, smarter electric grid that will allow for the broader use of alternative energy.

It is now clear that candidates for public office, advocacy organizations, business leaders, and elected officials ignore climate change and energy alternatives to their great peril. The saliency of these issues and their long-term impact on public opinion, election outcomes, and energy policy—particularly in the Intermountain West—has just barely begun to be felt.

Jill Hanauer is president of Project New West, a western research and strategy company based in Colorado. Hanauer has over twenty-five years of political experience. She has worked for US senators Gary Hart, Tom Harkin, and Barbara Boxer, and the Democratic National Committee under Chairmen Paul Kirk and Ron Brown. Hanauer was the Pro-Choice America director of the National Abortion and Reproductive Rights Action League and vice president of the Center for National Policy. She cofounded The Interfaith Alliance and served as its executive director for five years.

A lifelong Coloradan, **David Winkler** joined Project New West at its inception in 2006 and currently serves as research director. He has extensive experience with political campaigns, issue advocacy, civic engagement, and community organizing throughout the West.

Lisa Grove, president of Grove Insight, has conducted polls and moderated hundreds of focus groups and dial test sessions for presidential, senatorial, and gubernatorial candidates; labor unions; conservation groups; and other nonprofits, governments, and corporations specializing in the West, where she lives and plays. Grove is also the owner of two other environmentally responsible businesses, IF Green and Green Insight.

Melissa Chernaik is a senior analyst at Grove Insight, based in Portland, Oregon. Over the years, Chernaik has fulfilled myriad roles in the political world, including pollster, campaign manager, communications and field director, and general, direct mail, and message consultant. She also serves on several local boards of directors and conducts training for first-time candidates and staffers.

Andrew Myers is president and chief executive officer of Myers Research, based in Virginia. His clients include over 200 elected officials and a host of progressive organizations and think tanks. Myers has worked on elections at all levels, from local to presidential, and has served as the principal pollster to the Democratic Legislative Campaign Committee and the Democratic Governors Association. He has extensive experience in the West, particularly in Colorado, Montana, and Idaho.

HOMEGROWN SECURITY
By Chip Ward

Now that we've decided to green the economy, why not green homeland security, too? I'm not talking about interrogators questioning suspects under the glow of compact fluorescent lightbulbs, or cops wearing recycled Kevlar and recharging their Tasers via solar panels. What I mean is shouldn't we finally start rethinking the very notion of homeland security on a sinking planet?

Now that Dennis Blair, the new director of National Intelligence, claims that global insecurity is more of a danger to us than terrorism, isn't it time to release the idea of security from its top-down, business-as-usual, terrorism-oriented shackles? Isn't it, in fact, time for the Barack Obama administration to begin building security we can believe in—that is, a bottom-up movement that will start us down the road to the kind of resilient American communities that could effectively recover from the disasters, man-made or natural (if there's still a difference), that will surely characterize this emerging age of financial and climate chaos? In the long run, if we don't start pursuing security that actually focuses on the foremost challenges of our moment, that emphasizes recovery rather than what passes for "defense," that builds communities rather than just more SWAT teams, we're in trouble.

Today, "homeland security" and the Department of Homeland Security (DHS), that unwieldy amalgam of thirteen agencies created by the George W. Bush administration in 2002, continue to express the potent, all-encompassing fears and assumptions of our last president's Global War on Terror. Foreign enemies may indeed be plotting to attack us, but, believe it or not (and increasing numbers of people, watching their homes, money, and jobs melt away, are coming to believe it), that's probably neither the worst nor the most dangerous thing in store for us.

Outsized fear of terrorism and what it can accomplish, stoked by the apocalyptic look of the attacks of 9/11, masked the agenda of officials who were all too ready to suppress challenges by shredding our civil liberties. That agenda has been driven by a legion of privateers, selling everything from gas masks to biometric identification systems, who would loot the public treasury in the name of patriotism. Like so many bad trips of the Bush years, homeland security was run down the wrong tracks from the beginning—as the arrival of that distinctly un-American word *homeland* so clearly signaled—and it has, not surprisingly, carried us in the wrong direction ever since.

In that context, it's worth remembering that after 9/11 came Hurricane Katrina, epic droughts and wildfires, biblical-level floods, and then, of course, economic meltdown. Despite widespread fears here, the likelihood that most of us will experience a terrorist attack is slim indeed; on the other hand, it's a sure bet that disruptions to our far-flung supply lines for food, water, and energy will affect us all in the decades ahead. Nature, after all, is loaded with disturbances like droughts (growing ever more intense thanks to global climate change) that resonate through the human realm as famines, migrations, civil wars, failed states, and eventually warlords and pirates.

Even if these seem to you like nature's version of terrorism, you can't prevent a monster storm or a killer drought by arresting it at the border or caging it before it strikes. That's why a new green version of security should concentrate our energies and resources on recovery from disasters at least as much as our defense against them—and not recovery as delivered by distant, fumbling Federal Emergency Management Agency (FEMA) officials either. The fact is that preorganized, homegrown (rather than homeland) networks of citizens who have planned and prepared together to meet basic needs and to aid one another in times of trouble will be better able to bounce back from the sorts of disasters that might actually hit us than a nation of helpless individuals waiting to be rescued or protected.

Imagine redubbing the DHS the Department of Homegrown Security and at least you have a place to begin.

Homegrown Security for a Cantankerous Future

Homeland security post-9/11 has been highly militarized and focused primarily on single-event disasters, like attacks or accidents, not on, say, the infection of critical grain crops by some newly evolved disease or, as is actually happening, the serial collapse of ocean fisheries. Unlike a terrorist attack, such disasters could strike everywhere at once, rendering single-point plans useless. If Denver goes down in an earthquake, FEMA can (we hope) feed people via trucks and airdrops. If some part of the global food trade were to shut down, hundreds of thousands of community gardens and networks of backyard farmers ready to share their harvests, not warehouses full of emergency provisions, could prove the difference between crisis and catastrophe. Systemic challenges, after all, require systemic responses.

Food and security may not be a twosome that comes quickly to mind, but experts know that our food supply is particularly vulnerable. We're familiar with the hardships that follow spikes in the price of gas or the freezing of credit lines, but few of us in the United States have experienced the panic and privation of a broken food chain—so far. That's going to change in the decades ahead. Count on it, even if it seems as unlikely today as, for most of us, an economic meltdown did just one short year ago.

Our industrialized and globalized food production and distribution system is a wonder, bringing us exotic eats from distant places at mostly affordable prices. Those mangos from Mexico and kiwis from New Zealand are certainly a treat, but the understandable pleasure we take in them hides a great risk. If you're thinking about what the greening of homeland security might actually mean, look no farther than our food supply.

The typical American meal travels, on average, 1,000 miles to get to your plate. This is especially true in our western heartland, far from coastal ports. The wheat in your burger bun may be from Canada, the beef from Argentina, and the tomato from Chile. Food shipped from that far away is vulnerable to all sorts of disruptions—a calamitous storm that hits a food-growing center; spikes in the price of fuel for fertilizer, farm machinery, and trucking; internecine strife or regional wars that shut down harvests or block trade routes; national policies to hoard food as prices spike or scarcities set in; not to speak of the usual droughts, floods, and crop failures that have always plagued humankind and are intensifying in a globally warming world.

An interruption of food supplies from afar is only tolerable if we've planned ahead and so can fill in with locally grown food. Sadly, for those of us who live outside of California and Florida, local food remains seasonal, limited, and anything but diverse. In the West, we'd realize that a mostly meat and potatoes diet is as monotonous as it is inadequate. And don't forget, local food has been weakened in this country by the reasonably thorough job we've done of wiping out all those less-than-superprofitable family farms. United States agriculture is now strikingly consolidated into massive, industrial-style operations. So chickens come from vast chicken farms in Arkansas, hogs from humongous hog outfits in Georgia, corn from the mono-crop midwestern corn belt, and so on.

Such monolithic enterprises may be profitable for Big Ag, but they're not going to do us much good, given the cantankerous future already inching its way toward us. When a severe drought in Australia led to plummeting rice production in the Murray River basin last year, the price of rice across the planet suddenly doubled. The spike in rice prices, like the sudden leap in the cost of wheat, soy, and other staples, was primarily due to the then-soaring price of oil for farm machinery, fertilizer, and transport, though rampant market speculation contributed as well. At that moment, the collapse of Australian rice farming pushed a worsening situation across a threshold into crisis territory. Because the world agricultural trade system is so thoroughly interconnected and interdependent, a shock on one part of the planet can resonate far and wide—just as (we've learned to our dismay) can happen in financial markets.

Think of the shortages and ensuing food riots in thirty countries across the planet in 2008 as grim coming attractions for life on a planet with unpredictable extreme weather, booming populations, overloaded ecosystems, and distorted food economies. The spike in prices that put food staples out of reach of rioting masses of people was soon enough mitigated by the collapse of energy prices when the global economy tanked. Make no mistake, though: food shortages and the social unrest that goes with them will eventually return.

And here's something else to take into consideration: nations that suffer food shortages may, when their hungry citizens demand food sovereignty, protect their agricultural sectors by erecting trade barriers—just as is beginning to happen in other areas of production under the pressure of the global economic meltdown. The era of globalized food production, whose fruits (and vegetables) we Americans have come to consider little short of our supermarket birthright, may contract significantly in the relatively near future. We should be prepared. And that's where a Department of Homegrown Security could make some real sense.

Most American cities, after all, have less than a week's worth of food in their pipeline, and most of us don't stockpile, which makes city dwellers especially vulnerable to disruptions of the food supply. Skip your next three meals and you'll grasp the panic likely to arise if the American food chain is ever broken in a significant way. The question is, how can we address rather than ignore this vital, if underappreciated, aspect of homeland security?

Vertical Farms and Victory Gardens

Because cities are so dependent on daily food shipments, local food security in urban areas might well mean storing more food for emergencies. This would certainly be the old-school approach to disaster planning, and it has worked well enough over the short run. Over the long run, however, what makes real sense is to encourage urban and suburban community gardens and farmers' markets, and not just on a scale that ensures a summer supply of arugula and fresh tomatoes, but on one that might actually help mitigate prolonged food disruptions. There are enough vacant lots, backyards, and rooftops to host many thousands of gardens, either created by voluntary groups or by small-scale entrepreneurs. Urban farming could even go big. Columbia University professor Dickson Despommier recently unveiled his vision of a vertical farm, a thirty-story tower right in the middle of an urban landscape that could grow enough food to feed 50,000 people in the surrounding neighborhood.

Cultural historian and visionary critic Mike Davis has already wondered why our approach to homeland security doesn't draw from the example of victory gardens during World War II. In 1943, just two years into the war, 20 million victory gardens were producing a staggering 30 to 40 percent of the nation's vegetables. Thousands of abandoned urban lots were being cleared and planted by tenement neighbors working together. The Office of Civilian Defense encouraged and empowered such projects, but the phenomenon was also self-organizing because citizens on the home front wanted to participate, and home gardening was, after all, a delicious way to be patriotic.

Rebecca Solnit, author of *Hope in the Dark*, reports that, within the deindustrialized ruins of Detroit, a landscape she describes as "not quite postapocalyptic but...post-American," people are homesteading abandoned lots, growing their own produce, raising farm animals, and planting orchards. In that depopulated city, some have been clawing (or perhaps hoeing) their way back to a semblance of food security. They have done so because they had to, and their reward has been harvests that would be the envy of any organic farmer. The catastrophe that is Detroit didn't happen with a Hurricane Katrina–style bang, but as a slow, grinding bust—and a possibly haunting preview of what many American municipalities may experience postcrash. Solnit claims, however, that the greening of Detroit under the pressure of economic adversity is not just a strategy for survival, but a possible path to renewal. It's also a living guidebook to possibilities for our new Department of Homegrown Security when it considers where it might most advantageously put some of its financial muscle while creating a more secure—and resilient—America.

As chef and author Alice Waters has demonstrated so practically, schools can start "edible schoolyard" gardens that cut lunch-program costs, provide healthy foods for students, and teach the principles of ecology. The food-growing skills and knowledge that many of our great-grandparents took for granted growing up in a more rural America have long since been lost in our migration into cities and suburbs. Relearning those lost arts could be a key to survival if the trucks stop arriving at the Big Box down the street.

The present Department of Homeland Security has produced reams of literature on detecting and handling chemical weapons and managing casualties after terrorist attacks. Fine, we needed to know that. Now, how about some instructive materials on composting soil, rotating crops to control pests and restore soil nutrients, and canning and drying all that seasonal bounty so it can be eaten next winter?

It's not just about increasing the local food supply, of course. Community gardens provide a safe place for neighbors to cooperate, socialize, bond, share, celebrate, and learn from one another. The self-reliant networks that are created when citizens engage in such projects can be activated in an emergency. The capacity of a community to self-organize can be critically important when a crisis is confronted. Such collective efforts have been called *community greening* or *civic ecology*, but the traditional name *grassroots democracy* fits no less well.

Ideally, the greening of homeland security would mean more than pamphlets on planting, but would provide actual seed money—and not just for seeds either, but for building greenhouses, distributing tools, and starting farmers' markets where growers and consumers can connect. How about raiding the Department of Homeland Security's gluttonous budget for "homegrown" grants to communities that want to get started?

Here's the interesting thing: without federal aid or direction, the first glimmer of a green approach to homeland security is already appearing. It goes by the moniker *relocalization*, and if that's a bit of an awkward mouthful for you, it really means that your most basic security is in the hands not of distant officials in Washington but of neighbors who believe that self-reliance is safer than dependence. In this emerging age of chaos, pooled resources and coordinated responses will, this new movement believes, be more effective than thousands of individuals breaking out their survival kits alone or waiting for the helicopters to land.

Actually, relocalization is an international movement and, as usual when it comes to the greening of modern society, the Europeans are way ahead of us. There are now hundreds of local groups in at least a dozen countries that are convening local meetings as part of the Relocalization Network to "make other arrangements for the post-carbon future" of their communities. In Great Britain, an allied Transition Towns movement has sprung up in an effort to spark ideas about, and focus energies on, how to wean whole communities off imported energy, food, and material goods. With a rising sea at its front door, the Netherlands has taken a further step. Its national security plan actually makes sustainability and environmental recovery key priorities.

In the United States, post-carbon working groups are beginning to sprout across the country. In my backyard, right in the heart of red-state Utah, a diverse group of citizens calling themselves the Canyonlands Sustainable Solutions have come together to generate practical plans for insulating the remote town of Moab, 200 miles from the trade and transport hub of Salt Lake City, from future food and energy price shocks and supply interruptions. Such local groups are often loosely allied with one another, especially regionally, through websites and blogs that report on the progress of diverse projects, trade ideas as well as information, and offer lots of feedback.

The citizens engaged in relocalization projects have largely given up on federal aid and are going it alone. Still, think how much farther they could go if only a fraction of the $27 billion directed at state and local governments to enhance "emergency preparedness" in the 2009 Department of Homeland Security budget were given in grants to their projects. If we can afford to hand rural Craighead County in Arkansas $600,000 for hazmat suits and other antiterror paraphernalia to defend cotton and soybean farmers from attack, surely we could provide grants for urban homesteaders in Detroit.

Food security, of course, is just one aspect of a green vision of home-grown security. Other obvious elements like energy and water security could also be reimagined, if only official Washington weren't so stuck in the obvious. No doubt, somewhere out there on the *Titanic* this planet is becoming, the go-it-aloners, with no Department of Homegrown Security to back them, are already doing so—and helping prepare us all as best they can for the realization that, right now, there are not enough lifeboats to carry us to safety.

Perhaps it's not so unrealistic to expect that someday, as a homegrown security movement builds and matures, it can capture a share of the federal funds that now go to such dubious measures as closed-circuit televisions and crash-proof barriers at sports stadiums, including $345,000 for Razor-back Stadium in Arkansas.

In the meanwhile, let's encourage projects that are building resilience in communities as small as Moab and as large as New York City, while revitalizing local culture with a dose of grassroots engagement. Seed it and feed it, and it will bloom. Along the way we will learn that when it comes to home, or land, or security, living in an open, inclusive, and robust democracy is not an impediment to defense but a deep advantage. Democracy, if only we nurture it, is the very soil of our resilience.

Chip Ward is a former grassroots organizer who has led several successful campaigns to make polluters accountable. The author of *Canaries on the Rim* and *Hope's Horizon*, he writes from Torrey, Utah.

RETOOL FOR YOUR NEXT MISSION
By Florence Williams

An evening after St. Patrick's Day, the Celtic Tavern in downtown Denver is dead. A waitress listlessly polishes empty tables and a bartender talks on the phone. Garett Reppenhagen is hanging out with his buddies Ray Curry and Jeff Heckle at a table near the window. Reppenhagen's driven in from Colorado Springs for some meetings related to his new job as the state director of Veterans Green Jobs. He's thirty-three and sports a buzz cut, baggy jeans, and a blue button-down shirt. Curry's been living on Reppenhagen's couch and has come along for the ride. He is twenty-four, thin, and wearing loose black clothes. Heckle is now a grad student in anthropology in Denver. With long hair and a paunch, he looks like he is actively shedding his previous identity as an army specialist.

Reppenhagen and Heckle used to work together at the Home Depot in Grand Junction. Reppenhagen was a high school dropout who'd moved around a lot as a kid. Grand Junction was as good a place as any to land, and he had a girlfriend there. "We were young punk kids," he says. When Reppenhagen was twenty-five, they decided to enlist in the army to have some fun and see the world. It was one month before 9/11.

After basic training in Kentucky, Reppenhagen and Heckle shipped out to Germany, and then to Kosovo for nine months of peacekeeping as cavalry scouts. "We were like cops," says Reppenhagen, who was trained as a sniper and did things like deliver medical supplies and guard weddings on the border of Macedonia. When they got transferred to Iraq in January 2004, it was supposed to be more of the same. "Hussein had just been found and there were no weapons of mass destruction. It was going to be like babysitting the town," he says.

Only, as everyone knows, it didn't quite work out that way. Reppenhagen was part of a scout platoon in the 263rd Armor Battalion. The scouts were supposed to do recon and screening, but few of them had trained for small-unit tactics. Suddenly, they found themselves working in counter-mortar operations or as convoy escorts. They kicked down doors in hostile neighborhoods in Baqubah. North of Baghdad, just outside the Sunni Triangle, the town and surrounding Diyala River Valley were housing Abu Musab al-Zarqawi and became one of the strongest insurgent guerilla areas in Iraq.

One night, Reppenhagen was on a sniper mission on the roof of the Hib Hib police station. It was hot, so several police officers were sleeping up there. Early in the morning, the sniper team was swatting bugs and trying to stay awake when three vehicles drove by, stopping in a grove of palms by a wide canal down the street. Moments later, rocket-propelled grenades fizzed into the sky and arced onto the roof. "Pieces of bed frames scattered across the building," recalled Reppenhagen. The grenades kept coming and men screamed. The snipers' gunfire and grenades couldn't reach the palm grove, but Reppenhagen had mapped the grove's coordinates the night before. He called in a target grid to the mortar team on his base and waited. "The police dragged the wounded and dead down the concrete stairs into the building, so we were the only ones topside to see the mortars rain down onto the enemy's position."

Reppenhagen received an Army Commendation Medal for saving lives and for his quick thinking under fire. But despite his personal successes, the war was a huge disappointment. "I felt more like an occupier than a peacekeeper. It gave me a sour taste of the mission. Shock and awe was undermining winning hearts and minds. There was not a single soldier in Normandy in World War II who didn't think he was stopping the spread of fascism. This wasn't like that."

And coming home after a year in Iraq wasn't any easier. There weren't many jobs. There was insufficient care for his injured and traumatized friends; no one seemed to be interested in helping the veterans stay in college or train for work or keep their frayed family relationships together. "You do sixteen weeks of basic training to go to war, then you get to do a couple of days of training in resume writing when you get home?" Reppenhagen asks. "That's it?"

Reppenhagen started working for a Washington, DC–based advocacy group called Veterans for America, helping with public relations and shepherding legislation through Congress. But bill after bill failed, things that would help vets with brain injuries, and better versions of the GI Bill. He was angry and deluded. He volunteered for another Washington, DC–based group, called Iraq Veterans against the War, and that's where he met Curry.

Both Reppenhagen and Curry felt the war was about oil. And that, they decided, was a dumb reason to risk your life. The more they talked about it and thought about it, the more they wanted to do something. In March 2006, Reppenhagen joined a five-day antiwar and lame-government-response-to-Katrina march from Mobile, Alabama, to New Orleans. In Slidell, Louisiana, he stayed on the bayou property of a veteran named Gordon Soderberg, who was retrofitting buses and construction equipment to run on biodiesel made from used vegetable oil. Why fight wars over Persian Gulf oil when you can make your own? For Reppenhagen, the argument was a revelation. He hadn't been interested in environmental causes before, although he believed that his father, who had fought in Vietnam and died when Reppenhagen was thirteen, had died of Agent Orange–related cancer. He thought there should be a more sustainable way to live. "It made sense to me," says Reppenhagen. "Reducing dependence on foreign oil seemed like a solution to better national security."

Soderberg told Reppenhagen about a new organization called Veterans Green Jobs that was hoping to prepare veterans for green jobs. "My original intentions were just to help out, do outreach within the veteran community and with my contacts," says Reppenhagen. "I thought it was a great idea. I picked up more and more work with it, and soon I was an employee."

Veterans Green Jobs was a Colorado-based nonprofit hatching plans to train veterans in things like biodiesel conversion, making homes more energy efficient, and forestry conservation work like trail building and salvage operation. This was long before there was a new administration or a stimulus bill.

Reppenhagen told Curry about the program and signed him up for the first class of training, which started in just a couple of weeks. Reppenhagen goes around talking to vets' groups at the Veterans Administration and other places, and Curry usually comes along.

Curry feels the same way about fighting a war for the wrong reasons. "I didn't sign up to be a gas man," he says. Curry, who says today's root problem is the importation and overuse of resources, is a vegetarian. He doesn't drive a car or buy a lot of stuff. "I like to weigh my footprint as much as possible. It's a personal journey for me."

But it wasn't just the green-resource side of the jobs program that appealed to Reppenhagen and Curry; it was the people side. "A good, meaningful job and camaraderie and support are what a lot of veterans need," says Reppenhagen. And they need it soon after leaving the military, before the destructive cycle of substance abuse and depression sets in. "It's heartbreaking," he says. "Most of my friends with families are divorced now. The divorce rate is really high. You can't settle down or keep solid friends. You move around. These guys have post-traumatic stress that doesn't come in clichéd flashbacks," continues Reppenhagen. "Mostly, they just sort of become jerks. Their families kick them out, and they end up on someone's couch playing World of Warcraft, taking their meds and ordering pizza."

Curry, a wiry, intense guy, had been flailing around since his discharge from the marines in 2005. He'd done some humanitarian relief work after Hurricane Katrina, but mostly he'd been bartending around the DC area. "I went through three years of self-destructive cycles," he says. "A lot of veterans struggle in civilian jobs. It doesn't always pan out. I had authority struggles, some post-traumatic stress. I went from being a leader in the marines to working in bars in subservient and pointless jobs. It pays the bills, but it's not fulfilling." Some veterans have a hard time sitting in classrooms of students they can't relate to and they often drop out of college; others end up depressed, homeless, or suicidal.

"With the veterans these days, it's a typical story," agrees Curry. "They believe in service. That's why they enlisted. But they aren't really serving these days. They need to receive services. This is a way of employing us and getting into career paths." Nearly 2 million veterans have returned or are returning from Iraq and Afghanistan.

Curry is one of fourteen male recruits for this first training class (a women's training session is in the works). Joining him will be four formerly homeless veterans coming out of a program run by the Denver Department of Human Services. After a team-building camping trip in Utah, they will head off to Colorado's San Luis Valley, where they will live for eight weeks in the late spring in migrant-worker housing. In this unit, they will learn about retrofitting homes for energy efficiency, but also leadership, since these fourteen will help lead training programs to come in Louisiana, New Mexico, and Washington. Other training programs will include solar installation, a biodiesel vehicle-conversion project, first responder disaster preparedness, and forest conservation projects. By the end of the summer, 200 veterans will receive green-jobs training, with a goal of tripling that number in 2010. Trainees can receive college credit if they want it, and many will gain professional certification in at least one of four areas having to do with energy-efficiency retrofits: as a building analyst, "envelope" professional, home energy rating auditor, or computer systems auditing analyst.

During the first eight-week session, the men will gain certification in home energy audit training. They will practice their new skills—things like beefing up insulation; tightening windows; replacing water heaters, shower-heads, and lightbulbs on low-income houses, and they get paid for it. Housing and meals are also to be provided. If it sounds like boot camp, that's part of the idea. These men are comfortable living and working as a team, and they're used to hands-on training. Support and mental health services will be available if they need it. "They find self and purpose in the world and reconnect with their cohort group," says Reppenhagen. "We're empowering veterans to do something amazing in their lives where there's a sense of meaning and purpose again and fold that into beneficial work."

Veterans Green Jobs was the brainchild of Brett KenCairn, a Wyoming-bred community organizer who worked to retrain loggers in the Pacific Northwest after the spotted owl controversy and then worked with Native Americans on sustainable-forestry projects in Arizona and New Mexico. Now based in Boulder, he saw linking veterans and green jobs as solving two big problems: underemployed vets and an existing workforce too small to tackle global warming. "If we don't figure out how to mobilize a new workforce at a dramatic scale, our chances of averting climate change are virtually nil," says KenCairn, who never fought a war but whose father served in Vietnam. "We need to retrofit every building in our built environment. Veterans represent one of the best workforce assets because they're already ready for rapid training and deployment."

Walmart loved the idea and provided $750,000 in seed money for the first trainings. General Wesley Clark and former Central Intelligence Agency director James Woolsey have signed up for the advisory board. KenCairn's group will likely get a boost from the 2007 energy bill and even more from Obama's stimulus bill, the American Recovery and Reinvestment Act of 2009. It directs about $50 billion toward renewable-energy and efficiency programs nationwide, including about $130 million for weatherizing homes and $50 million in other energy programs in Colorado alone. In the state, that could lead to the creation of 59,000 jobs, and $3 to $4 million nationally. KenCairn says Veterans Green Jobs will bid to be a contractor for state agencies newly enriched by the stimulus money. He hopes the trained veterans will be well positioned for new jobs at Xcel Energy and at wind and solar companies. "We can train rapidly," says KenCairn. "This is a perfect entry point for weatherization and other green job pathways. Our motto is 'Retool for Your Next Mission.'"

In Colorado, the Governor's Energy Office is excited by the idea. Currently, the office funds and oversees weatherizing of about 4,000 homes a year. The stimulus bill will double or triple that for the next three years, says Deputy Director Seth Portner. The work will be bid out, and Portner thinks Veterans Green Jobs stands a good chance of winning the work. "Having been acquainted with the Veterans Green Jobs concept, I think it's nothing short of brilliant," he says.

So does Steve Byers, a Gulf War veteran who started EnergyLogic, a thirty-employee energy-systems company that works along the Front Range. Byers led some of the training for the veterans during the first spring and summer. "One of the things about this field is that there's a mission involved," says Byers. "We're used to an overarching mission in our lives. This work is a good fit. It's not brain surgery, but it requires technical training, and that mirrors the military. I'm learning something new almost every day. And it has an obvious resonance after fighting a war that had something to do with petroleum resources. That's why I got into the field."

Both Reppenhagen and KenCairn expect the military may just be the bridge America needs to popularize the green economy. "The link to average Americans is missing right now," says Reppenhagen. It's one thing to want to be a hippie in Boulder and hug a tree; it's a whole other level to be a veteran and say, 'Hey, I'm coming home from [a] war fighting for oil.' I think the culture clash could be decreased by realizing there's something seriously patriotic about energy independence."

Florence Williams is a contributing editor at *Outside* magazine. She also frequently writes on science and the environment for *The New York Times*, *OnEarth*, *High Country News*, and other publications. A former Ted Scripps Fellow in Environmental Journalism at the University of Colorado, she lives in Colorado with her family.

LESSONS FROM ACROSS THE POND
By Heidi VanGenderen

"Climate change is like the Internet. It's getting bigger every year, it's not going away, and you need to figure out how to make money from it."

This entrepreneurial observation comes from Paul Dickinson, cofounder and executive director of the Carbon Disclosure Project (CDP) in London. The CDP is a nonprofit organization that has amassed the "largest database of corporate climate-change information in the world." The CDP issues questionnaires to companies of all stripes, and the responses can be accessed by essentially anyone, including institutional investors, governments, and individuals. In a world in danger of being filled with too much information, CDP's database is information that can matter, because it can and does help guide investment decisions. And that is one key strategy in our ability to meet both the challenges and the opportunities of climate change.

We who are fortunate enough to reside in the western United States know challenge well. We live on unforgiving lands, and, by god, we can face adversity and come through whatever comes at us because we are self-reliant. We are rugged individuals who also can demonstrate an entrepreneurial flair. When it comes to climate change, can and will we apply both of these traits? Will we do so, in part, by reaching beyond the borders of our own well-loved region to learn what is working elsewhere so that we may borrow and adapt good ideas?

Awareness and acceptance of the human ability to alter the Earth's atmosphere clearly differ around the planet. I know this better now through the opportunity to gather in an international fellowship thanks to the British Foreign and Commonwealth Office. Fourteen of us representing eleven different countries listened, learned, and debated solutions to climate change through the lens of finance and investment.

The perspectives from colleagues, and now friends, from China, India, Malaysia, Brazil, Kazakhstan, Turkey, South Africa, Canada, Australia, and Indonesia vastly expanded my personal horizons. Because we were based at the University of Edinburgh, with several forays to London, one of the world's premier finance capitals, we were treated, in particular and understandably, to a UK perspective.

For starters, the public presence of climate change is greater in the United Kingdom than in the US and most certainly far greater than, say, in Kazakhstan. Whether on the British Broadcasting Corporation television news in the debate over whether another runway should be added at Heathrow, in product labels, in public displays in museums, or in ads everywhere, the direct topic of climate change and the imperative to reduce carbon emissions is a much greater part of the UK's lexicon than it is of the US one. We in America are more likely, perhaps, to use the phrase *global warming* than in the UK, as they seem to use *climate change* nearly exclusively—but we are far less direct in our address of the topic overall than those in the UK.

Consumer labeling is an evident strategy in the UK. The supermarket chain Tesco, for example, has taken the initiative to list the amount of carbon dioxide emitted in the production and distribution of some of the products it sells so that concerned consumers can use this information in their purchasing decisions.

All new vehicles in the UK are labeled with grams per kilometer (g/km) on carbon emissions, and under new law, the highest-emitting cars (225 g/km) will be taxed an extra £950 (roughly $1,500) just to drive off the showroom floor. Consumers and citizens are hard-pressed to miss the radio ads, print ads, and community outreach by a multitude of public, private, and nonprofit entities asking for their awareness and action to "lower your carbon footprint."

Of course, consumer labeling assumes that consumers give a hoot about climate change and carbon footprints, so while the verdict isn't in yet regarding the effectiveness of the effort, it is at least there for the taking.

We visited several different kinds of energy-production facilities while in the UK, including a municipal combined heat and power system in an oil industry boomtown, a nuclear plant, a waste-to-energy facility, and a woody biomass plant. The latter made me think about my native land in the western US in particular.

The Steven's Croft biomass plant in Lockerbie, Scotland, produces fifty megawatts of electricity from wood chips made from farmed trees and "waste" wood like doors and construction waste that would otherwise go into a landfill. The plant burns half a million tons of wood each year to provide the power needs for 70,000 homes. Its feedstock is and will remain near the plant, which went into production two years after the groundbreaking.

There is currently limited use of the plethora of "red dead" trees in the western US for power production or any other use. Distances are far greater, but the challenge of those trees only grows.

There are small signs that the entrepreneurial spirit of the West is starting to come forward regarding these trees. One example is the recent opening of the Blue Wood Showroom in Breckenridge, Colorado, which showcases products and services using lumber from beetle-kill trees.

The visitors' center of the biomass facility in Scotland is covered with displays and posters with titles like "Tackling Climate Change Is at the Heart of Everything We Do," "How We're Fighting Climate Change," and "How Climate Change Is Affecting Us All." Can the Blue Wood Showroom use the opportunity to do something similar?

There is an enormous array of daily printed newspapers in the UK where the price of newsprint must not be as high given the size of most of them. US newspapers use significantly less paper. In the UK, however, almost without exception, there is at least one article each day that references carbon emissions in most of these publications. For example, a story on the Scottish Green Buildings Plan reads as follows: "New buildings must be more energy efficient next year, the Scottish government said. Carbon emissions must be down by 30 per cent on those from existing buildings…" Advertisements in these publications and elsewhere frequently tout a carbon attribute. As only one among many, many examples, take the National Express coach service ad that notes, "Carbon Footprint: Travel Shouldn't Cost the Earth." (Best of luck with that one?!)

All cynicism aside, the critical importance of intergenerational equity and education is also evident in the UK. We delighted in playing an interactive game intended for kids at the National Museum of Scotland. In it, you are made the energy czar of the country and charged with building a low-carbon future by selecting power-production options that differ geographically, socially, and economically. The whole of the game is repeatedly framed within the imperative to reduce greenhouse gas emissions, and the measurements of those reductions are also clear throughout.

It's not that those in the UK are without an attempted sense of humor about the whole dilemma in their greater diligence toward casting a light on it all. While shopping for a birthday card, I came across a greeting card that shows a distraught young woman kneeling in a medieval palace kitchen next to her scrub bucket with a caption that reads, "Cinderella couldn't help but worry about her carbon footprint." I, of course, completely identify with Cinderella in so many ways…

The differences in recent political leadership in our respective countries are hugely evident when it comes to climate change. Policies enacted in the UK can be credited as at least one key ingredient within this heightened presence. There is a fairly impressive array of policy instruments in the UK that more overtly address the goal of greenhouse gas reductions and use that metric more than anything yet in the US.

The UK has passed legislation that introduces the world's first long-term legally binding framework to tackle the dangers of climate change in its Climate Change Act of 2008. The 65 percent reduction noted in this legislation, now law, may be upped by Scottish Parliament, which under the recently ascended majority of the Scottish National Party is debating legislation that would codify an 80 percent reduction target by 2050 for Scotland. One can argue whether setting targets is the most effective approach, but the UK has an internal competition going among its member countries on which can out-green the other. While this is starting to happen in the US in competition among states, it again isn't couched in terms of climate change, but rather more in the form of diversifying our energy portfolio, as in Colorado's much-touted New Energy Economy, for example.

The UK Climate Change Act included formation of the Committee on Climate Change, an independent body that will, among an array of tasks, make recommendations to the government regarding incremental carbon budgets that will help ensure meeting the longer-term reduction targets. Again, the metric of carbon-emission reductions is ubiquitous because, in part and as noted by Lord Adair Turner, "what gets measured gets managed." The Barack Obama administration might do well to take note of this Committee on Climate Change, as well as the applied metric of greenhouse gas emissions reductions.

Additional policies and programs in the UK include the Community Energy Savings Programme, the Climate Change Levy, the Industrial Emissions Directive, the Carbon Capture and Storage Initiative, Carbon Reduction Commitment, the Coal Forum, and the Environment Transformation Fund. As in the US, there are a number of policies in place on the energy and social sides of the ledger, including the Renewables Obligation, the Biomass Strategy, the Energy Technologies Institute, and the UK Fuel Poverty Strategy.

And, of course, there is the European Union's Emissions Trading Scheme (EU ETS), the largest multicountry, multisector greenhouse gas emission trading scheme in the world. A market-based mechanism established through policy, the EU ETS seeks to reduce greenhouse gas emissions through a cap-and-trade program. The EU ETS has been in existence since January 2005. The ironic twist is that it's based on the successful sulfur dioxide emissions trading system implemented in the US under the Clean Air Act Amendments of 1990. The story goes that in the international climate negotiations leading up to the Kyoto Protocol, the US vociferously objected to a carbon tax as an acceptable mechanism. So the European Union embraced the cap-and-trade concept and moved ahead with the EU ETS. The US, in the meanwhile, neither ratified Kyoto nor enacted a cap-and-trade or a carbon tax. We are nothing if not ironic.

Is direct address of climate change, as evidenced in the UK, more effective than addressing the energy economy, as is the preferred route thus far in the US? Is this an either/or? Not if the US embraces the concept of creating a prosperous market that creates profit in decarbonizing.

Fans of cap and trade, like James Cameron of Climate Change Capital in London, believe that the carbon market has already obtained "remarkable achievement" from a "remarkable experiment." Cameron notes that the market has a resolutely single purpose, the capacity to transcend national borders, and can be transformative. And that if we structure it well and right, it can be as big as the fossil-fuel market in twenty years' time. Significantly, this same market can serve as a primary instrument to infuse large amounts of capital into a low-carbon strategy. Finally, he wonders, is there a better instrument that anybody can think of?

Energy is politics. Tackling climate change is largely about energy and, therefore, also essentially a political and economic challenge. Sound science and education ideally create the platform for sound political and policy decisions that in turn help determine the economics.

In the absence of so many years of federal leadership in the United States, state and regional efforts have attempted to move this country forward. Now the breath of welcome fresh air known as the Obama administration stands a true chance of bringing the US back into a leadership position on the imperatives of transforming the global energy economy and consequent address of climate change.

Cameron also notes that we would be foolish to "waste a good crisis," and we in the West have always loved a good crisis. We have loved exploring new frontiers even more, and there are brave frontiers awaiting us. The business opportunities, whether directly in the carbon market or in sustainable endeavors that will both diversify the energy economy and create genuine credits for that market, abound on both sides of the pond.

Can and will the regional leadership around the world effectively learn from respective efforts in this globally (and yet not fully) connected world? Can the rugged individual entrepreneurs of the West in turn help lead the US? Here's hoping.

Heidi VanGenderen is a third-generation Colorado native who served as the state's first gubernatorial climate advisor. She has worked on energy and climate issues in the nongovernmental organization, public, academic, and private sectors and recently completed the Chevening Fellowship in Edinburgh and London.

JOURNALISM AND THE SCIENTIFIC CONSENSUS ON GLOBAL WARMING

By Jason Salzman

You'd expect a newspaper like *The Denver Post* to give major play to the story about mountain pine beetles devouring Colorado's lodgepole pines, and it is.

It's no Jon Benét Ramsey–style media frenzy, but the pine-beetle infestation was the focus of fourteen staff-written news articles from January 2008 through May 2009 in the *Post*, covering everything from its potential impact on tourism to legislative efforts to fund beetle-related battles.

But the *Post's* coverage of the possible connection between the dying forests and global warming has been skimpy at most.

The *Post* reported in a 2008 article that warmer winters are one of the factors, in addition to drought and forest maintenance, that seem to be causing the pine-beetle problem.[1] But the 1,850-word piece didn't mention climate change as a possible explanation for the warmer winters.

"I was the fill-in reporter, because we didn't have an environmental reporter at the time," *Post* reporter Howard Pankratz, who wrote the article mentioning warmer winters, told me. "I just wanted to report what they factually presented. I didn't even think about going into the global-warming aspect of it. You know, I think a lot of people could have read through the lines."

You have to go all the way back to 2005 to find the phrase *climate change* or *global warming* in a *Post* news story focusing on pine beetles. (This tally does not include *opinion* articles, like editorials, but only staff-written *news* stories.)

The 2005 news story offered this paragraph:

"The system of checks and balances is a little out of whack," says veteran entomologist Dave Leatherman of Fort Collins. "Because of climate change, whatever the cause of that is, beetles throughout the West are doing things people have never seen before."[2]

Since that paragraph was published in 2005, two *Post* news stories about the broad impacts of global warming briefly mentioned pine beetles. A May 2008 *Post* article about a federal study of the effects of climate change included a sentence stating that "a warmer climate will spread the range of pests, such as the pine bark beetle and the spruce beetle, as well as expand the range of invasive, nonnative grasses in the Southwestern deserts."[3] A January 2009 *Post* article about a new study referenced the beetles as one of possible climate-change impacts that may have caused the shortening of the life of trees in the West.[4]

I called *Post* environmental reporter Mark Jaffe to find out why his newspaper hasn't written more about the link between pine beetles and climate change.

"Pine-beetle infestations are normal, but this one is particularly bad possibly because the trees are stressed due to warmer temperatures, which may be a result of changing climate, and you don't get the winter knock back from an extreme temperature event," Jaffe told me. "You can't link [the pine-beetle infestation] directly to climate change," he added. "There are no scientific links. So all we've been able to get researchers to say is that these two elements, which may be making this infestation worse, could be related to the change in climate."

The *Post* has done this, but barely—with no stand-alone story addressing the topic, and just a few passing mentions, over at least the last three and a half years.

The *Post*'s news coverage about the pine beetles raises the question of whether journalists should discuss the possible role of global warming when reporting on an event that may—or may not—be caused by it. And if they do mention global warming in this context, are journalists obligated to quote skeptics who may *not* think global warming is occurring at all?

Addressing the first question, Christy George, special projects producer at Oregon Public Broadcasting, told me that when it comes to covering events like forest fires or hurricanes, reporters should explain the possible role of climate change in their stories. (George is the current president of the Society of Environmental Journalists, but spoke to me as an individual reporter, not on behalf of that organization.)

"It's not that we want bad science, where people say [a hurricane] is caused by climate change," she said. "But the good science that says you can't say this is climate change, but this is what we'd expect with climate change."

The pine-beetle story also deserves this type of journalistic treatment, with different views on the possible role of climate change in the infestation. And should global-warming skeptics be quoted?

George observed that the most hard-core global-warming skeptics have made a shift, previously asserting that there was no such thing as climate change at all but now saying the climate is changing, but humans are not responsible. She thinks the views of these skeptics need not be included in stories.

"There's no value to me as a reporter to continue to throw in that person who says humans aren't causing climate change at all, because we're just past that, in terms of the scientific evidence," she said. "There are tremendous disagreements about the impacts [of climate change] and what to do. We don't have to look hard to find conflict in the story."

I called Seth Borenstein, a science writer for the Associated Press, and asked to interview him about how reporters treat global-warming doubters at a time when there's a clear consensus among scientists globally that human-caused emissions are causing climate change.

He told me I needed to put my questions in writing, to get clearance.

I e-mailed him:

My very basic questions could be covered in a ten-minute interview, or, if you'd prefer, you can simply answer them via e-mail.

They are:

In general, at what point does a scientific consensus become so great that you feel less obligated to present the view of skeptics of that consensus?

Specifically, is the scientific consensus that humans are causing global warming now so overwhelming that you feel less obligated—or not obligated at all in some instances—to present the views of skeptics?

I can call you at any time you're available, if you want to do this by phone. Whatever's easier for you is fine.

Borenstein's response:

Jason,

I'm sorry, but my boss and corporate communications advise against me doing this. Sorry.

He suggested other reporters I might call, which was kind of him. But I couldn't resist e-mailing him that I thought it was way hypocritical for the Associated Press to refuse any legitimate interview, given that AP relies on interviews to keep its doors open.

But fortunately I had better luck with Andrew Revkin, a science writer at *The New York Times*. "The further you get away from the basic question, will human-generated greenhouse gases warm the world? to more specific ones, like, have we already measurably done that or what's going to happen with hurricanes? the shape of the curve of understanding can change very quickly," he said. "There's a lot more real dispute.

"What I try to do in stories like that is to remind people that there's an overall consensus on the basics that humans are now influencing climate, even as I explore the uncertainties related to some particular outcome," he added.

Revkin and I discussed a recent study showing that Antarctic ice shelves holding the glaciers from slipping into the sea are weakening. The findings were particularly significant because, as *The New York Times* reported in its article on the new Antarctic data, global-warming skeptics had long pointed to temperatures in the Antarctic, which appeared *not* to be warming, to question the reliability of computer-generated climate models used to predict planetary warming.[5]

Revkin pointed out that there's still lots of uncertainty about how global warming is affecting Antarctica.

"That particular question of the climate changes across Antarctica is still in the early stages of understanding: limited measurements, short time periods, not remotely similar to the question, are humans warming the world?" said Revkin.

On his *New York Times* Dot Earth blog, Revkin has tried to articulate the least disputed realities on climate by posting suppositions like more carbon dioxide means a warmer world. "It's hard to find serious dispute, but how warm?" he says. "There's a very wide range of possibilities. It's not just a range of views, but a range of outcomes—the sensitivity of the climate system to a doubling of carbon dioxide is hugely uncertain. It could be two degrees. It could be five degrees. That's the difference between manageable and disaster. And no one disputes that either, so that's not a function of differing views."

David Kopel, the research director at the Independence Institute, a Colorado-based free-market think tank, insists that there is no scientific consensus on global warming.

"The notion [of scientific consensus] is disinformation created by global-warming extremists to attempt to suppress the views of persons who believe that, to the extent that anthropogenic global warming exists, it is something for which people can readily make relatively minor adjustments," he e-mailed me.

"More broadly," he added, "true science is by its nature not a 'consensus' operation. Many important scientific discoveries have demonstrated the falsity of the then-current 'consensus' of scientists."

I asked Kopel, who is a former media critic for the defunct *Rocky Mountain News*, if he thought reporters should include skeptics in news stories claiming the Earth is round.

"Journalists should not feel obliged to present the view of people who believe that the Earth is flat, or that it is a cube," he wrote. "They should present the view of what you call 'skeptics' who can point to scientific evidence that it is more accurate to describe the Earth as an oblate spheroid, rather than 'round.'"

Good point. So Kopel agrees that consensus is possible, if not necessarily permanent. Journalists would agree with that too.

As Revkin told me, "Scientific understanding is a journey, not a single study, not a single IPCC [Intergovernmental Panel on Climate Change] report. The strength of the IPCC reports lies in the twenty-year history of the IPCC enterprise. It provides a directionality. If the fifth IPCC report starts to raise new questions about the strength of human influence on climate, that's important."

So, for now, just as journalists can ignore people who refuse to accept that the Earth is round (I mean, an oblate spheroid), they no longer need to cover those who dispute that humans are contributing to global warming—unless the doubters have something new to say and new evidence to prove it.

Reporters should assert as fact that global warming is happening and humans are contributing to it.

As for the pine beetles, they present journalists with an opportunity to write about the possible connection between a local problem and an international scientific debate. That's partially what regional journalism should be about. The beetles are part of the global-warming story, even if their population explosion turns out to have nothing to do whatsoever with climate change—or, more likely, we never know *for sure* because of the complexities involved.

Jason Salzman is an award-winning writer and media consultant. His articles and commentaries have been published in the *Bulletin of the Atomic Scientists*, *The Christian Science Monitor*, *The Chronicle of Philanthropy*, the Harvard *International Journal of Press/Politics*, the *Los Angeles Times*, *Newsweek*, *Nonprofit World*, *Sierra*, *Utne Reader*, and elsewhere. He's a former media critic for the *Rocky Mountain News*, and is the coauthor of *Making the News: A Guide for Activists and Nonprofits* and *50 Ways You Can Help Obama Change America*. Salzman is cofounder of Effect Communications.

Notes

1. Howard Pankratz, "'Catastrophic': The Beetle Infestation That Is Expected to Kill All of Colorado's Mature Lodgepole Pines Is Moving into Wyoming and the Front Range," *The Denver Post*, January 15, 2008.
2. Jack Cox, "Tiny Beetle Killing Millions of Pines," *The Denver Post*, September 13, 2005.
3. Mark Jaffe, "Climate Change's Deep Impact: Where Heat Meets West," *The Denver Post*, May 28, 2008.
4. Jennifer Brown, "Report: Warming Cuts Trees' Life in Half," *The Denver Post*, January 23, 2009.
5. Kenneth Chang, "Study Finds New Evidence of Warming in Antarctica," *The New York Times*, January 21, 2009.

GREEN-CITY LEADERSHIP
By Beth Conover

A subtle but significant shift began to take place in American local government around 2003. Environmental innovation, previously the realm of regulators and engineers, became a focal point and goal for elected officials, who also developed an interest in greening proactively—going above and beyond measures required by law. Suddenly green was hip in city government. Mayors began promoting solar power and living rooftops, and city councils embraced green buildings and the Kyoto Protocol. *Vanity Fair* magazine captured this trend when it profiled the greenest mayors on pages next to green movie stars and other personalities. Years before significant movement was seen at the federal level, local governments led a greening revolution in the United States.

In 2004, it was possible to count on two hands the number of major cities with staffed and funded sustainability initiatives at the executive level. By 2006, the number of funded programs had grown to include dozens of cities nationally. And by 2007, the trend had reached a fever pitch, with over 500 mayors signed on to the US Mayors' Climate Protection Agreement, committing to the spirit if not the letter of the Kyoto accords—a document that only a few short years earlier was not seen as safe political material by many. That number exceeded 900 in 2009.

What changed? How did cities, which lie at the bottom of the federal/state/local regulatory chain, come to lead a national trend in green government? And looking back, what has the green-city movement accomplished? Is this a genuine change of direction or just a passing trend? What makes a green-city program successful?

My Experience in Denver

My experience with these questions is firsthand. As a special assistant to Denver mayor John Hickenlooper from 2003 to 2004, I helped the mayor develop policy positions on issues related to parks, planning, public works, and water. In late 2004 and 2005, inspired by a conversation with Portland's sustainability chief Susan Anderson, I worked with Mayor Hickenlooper and Chief of Staff Michael Bennet to design and develop the mayor's Greenprint Denver program. I begged, borrowed, and stole ideas from a close group of peers in other cities across the country, all developing fledgling programs at a time when there was a collective sense of great new potential, as well as fierce competition driven by new national city rankings by groups like SustainLane and The Green Guide. From 2005 to 2007, I built Greenprint Denver into a citywide program and worked with city staff, scientists, and a high-level community advisory group to develop a climate action plan that aims to reduce the city's greenhouse gas emissions by 25 percent of 2005 levels by the year 2020.

Greenprint Denver is now among the largest initiatives in the mayor's office, with a permanent and borrowed staff of nine city employees and a combined annual budget of millions (in grants of state and federal dollars primarily). Solar America City grants, as well as stimulus funds, including new Energy Efficiency and Conservation Block Grants (EECBG) created by the Barack Obama administration, have helped fuel a new generation of related programs at the city level at a time when they are badly needed.

And in 2009, Denver received the US Conference of Mayors' Climate Protection Award for its work on the regional FasTracks transit initiative—the most ambitious transit initiative in US history, connecting smart growth, housing choices, and expanded transportation regionally. Denver also took the top award from the US Chamber of Commerce Business Civic Leadership Center and Siemens Corporation in 2009 for its Greenprint Denver program.

What Changed?

Green-city programs have grown quickly in the West and nationally for a number of reasons. Perhaps first among them is a perceived livability edge to drive economic development. Communities recognize that success in reducing emissions and improving their environmental health makes them attractive to the multitude of creative-class workers who choose where to live based on reasons beyond job location, including the greenness of the community. The rise of foundation-sponsored healthy-community programs that link public health to community environmental health has also played a role by supporting a number of common goals, such as pedestrian and bicycle friendliness, urban gardens, and access to transit and good air and water quality.

The trend by cities in the past several years to track and measure performance data (as with Baltimore's CitiStat) and indicators of community health has also helped. Pioneering programs like Sustainable Seattle, started in the 1990s, recognized the importance of tracking and measuring multiple indicators beyond the traditional economic ones. The next generation of sustainable-city initiatives went beyond measuring indicators to developing policies to effect positive change. What gets measured gets done, and where cities have measured the impacts of city operations for per capita energy use, greenhouse gas emissions, and other key indicators over time, they have been likely to influence those variables.

Green-city programs are also being driven significantly by a shift in public interest and opinion. Such programs generally receive overwhelmingly positive responses from constituents, who reward the elected officials who champion them. Long before federal stimulus funding was tied to green jobs and energy efficiency, positive public opinion (and positive press) helped to boost the fortunes of many nascent green-city programs. Green programs almost always provide good news and great publicity opportunities, ideally with real impacts and content behind them.

The Evolution of Green-City Programs

When we set out to build Greenprint Denver, we did a quick national inventory of other strong green-city programs around the country and learned a few key lessons.

For example, many cities have demonstrated innovative efforts in one or two departments for years, like Denver's green fleet, Austin Energy's efficiency efforts, Baltimore's Urban Resources Initiative work on parks and water quality, and New York City's urban gardens. The successful city-greening programs that I describe here, in contrast, have sought to demonstrate environmental leadership in operations across departments at the highest level, to incorporate sustainability practices and values citywide, and to provide accountability metrics to back up practices and policies.

In 2004, some of the best city programs were those in Seattle, San Francisco, Chicago, Salt Lake City, Oakland, and Portland, Oregon. These programs evolved over different time frames and for different reasons—some through citizen initiative (as in San Francisco and Seattle), others by mayoral directive (as in Chicago and Salt Lake City). Some developed over a decade, others in a period of one to two years. All of them liberally and openly borrowed ideas from one another and benefited from a sense of competition (with cities regularly leapfrogging each other in the level and sophistication of their green-building policies, for example). Portland's office was earlier and larger (in terms of budget and staff) than others and benefited from competition with Seattle and San Francisco, other early leaders.

Chicago joined the fray later with a talented staff and a comprehensive set of actions (green rooftops, bike racks, and green affordable housing), borrowing its model from Seattle. Seattle raised the bar by initiating the US Mayors' Climate Protection Agreement, an initially modest effort organized for the 2005 ratification of the Kyoto Protocol that grew into a national phenomenon and demonstrated an American electorate at distinct odds with a national leadership that had refused to ratify that agreement. Salt Lake City and Oakland both focused on the economic benefits of city-greening programs and used mandates from their elected officials to build strong programs. Denver's Greenprint program, which started as one and a half people with no budget and a strong (volunteer-built) website, quickly emerged at the front of the second-generation programs.

Common Threads

In spite of their differences, a few common threads can be found among these diverse successful programs.

A STRONG EXISTING PROGRAM BASE

All of these cities had some existing pieces in place to support the broader vision and goal. (For example, in Denver these included strong programs in recycling; airport operations; environmental health; parks, trails, and natural areas; and a history of regional work on air-quality and transportation issues.) Existing programs make it easier to build a critical mass of interest within city governments and their populations and to provide clear and focused messaging.

POLITICAL WILL

These programs have a leading elected official in the role of champion. Who that person is depends in part on the form of city government. In strong-mayor cities, it is the mayor. In cities with strong councils and weaker mayor forms of government, it is often a leading councilperson in combination with the city manager. In counties, it is frequently one or more commissioners.

The champion is able to provide a clear priority message and direction to department directors to support the program through budget priorities, establish performance accountability measures, and so on. While champions within government are critical, champions from local nonprofits are also needed to keep the momentum and show community support for city officials during critical policy-making junctures.

FUNDING AND LEVERAGE

Strong city programs have access to funding, either via a dedicated budget or via budgeting authority over related departments, and through the support of community partners and philanthropies. Other city events and initiatives are engaged wherever possible to create new opportunities to showcase and advance the sustainability agenda—for example, in Denver the new downtown Justice Center became the first major city project to be built entirely to Leadership in Energy and Environmental Design (LEED) silver standards, and a recent bond issue generated competition among the city's cultural facilities to out-green one another in plans for new capital facilities. The hosting by Denver of the Democratic National Convention (DNC) in 2008 provided an unparalleled opportunity to engage businesses in citywide green-practices training and to attract new sponsorship dollars for programs like Freewheelin Denver, a public bike-sharing partnership that has evolved into a permanent program (Denver B-Cycle) to be launched in 2010, featuring up to 600 bikes at forty to fifty stations around the city.

COMMUNITY SUPPORT

Successful city programs enjoy community support, whether via grassroots initiative (as with the San Francisco Commission), appointed advisory group (as in Denver's Greenprint Council), or informal engagement. This support is often broadly representative of diverse populations, includes community opinion leaders, and is able to advance the legitimacy of the program in the broader community. Denver's advisory council support led to (among other things) development of a business-greening program by the Downtown Denver Partnership. Further partnerships with universities and state and federal agencies eventually helped in plans for the greening of the DNC in Denver in 2008.

Cities for Climate Action

So, are green cities a trend or here to stay? In a few short years, the green-cities movement has matured considerably. At the time of this writing, over 900 cities, representing over 80 million people, have signed on to the US Mayors' Climate Protection Agreement—forty-seven of these in the Rocky Mountain Region.[1] The movement has become more of a standard and is no longer quite as seat-of-the-pants. Cities have moved on in many cases from setting ambitious visions and goals to the real work of developing and implementing programs. With major federal money now in the game via EECBGs and State Energy Program funds, even cities that were not driven by climate goals are developing related programs in order to qualify for discretionary dollars.

In 2006–2007, Denver developed and adopted a citywide greenhouse gas inventory and Climate Action Plan with the help of the University of Colorado at Denver and the mayor's Greenprint Council.[2] The inventory describes per capita emissions of twenty-five metric tons of carbon dioxide per year for the city—about average for American cities (though above that of cities where coal is not the primary source of electrical power). The Climate Action Plan lays out a process to achieve community-wide reductions of greenhouse gas emissions by 10 percent below 2005 levels by 2012. This is equivalent to eliminating a small (250-megawatt) coal-fired power plant, or taking about 260,000 cars off the road. The plan further aims to decrease total community-wide emissions by 25 percent over 2005 levels by 2020. This is equivalent to eliminating two small coal-fired power plants (550-megawatt), or taking about 600,000 cars off the road.

The Denver plan provides a portrait of the challenges facing many medium-size cities that seek to reduce greenhouse gas emissions. About half of all emissions in Denver come from buildings and facilities (through electricity use), another third come from transportation (through tailpipe emissions), and the remaining come from waste and materials processing and management (including cement).

Emissions are the result of dozens of individual and public and private decisions made every day about building management, transportation choices, and material use and disposal. In seeking to reduce emissions, cities have relatively few policy levers to influence these decisions. Potential strategies for the Climate Action Plan were evaluated in terms of cost-effectiveness, potential impact, and their potential to engage the public. Final strategies combined public outreach and incentives with relatively minor policy interventions.

Denver's plan, like those in many leading cities, focuses on reducing emissions through three primary means: leading by example, outreach to private residential and business sectors, and the use of policy interventions in limited cases. Public outreach and incentives include the development of financing mechanisms for energy efficiency (through revolving loan funds, special districts, financing tied to property taxes or utility bills, and other means). Green-business training was developed in preparation for the DNC and continued through the city's Greener Denver Business Program, and neighborhood Energy Blitz weatherization efforts are now under way in low-income neighborhoods.

City lead-by-example programs range from the installation of major solar arrays on public buildings (at Denver International Airport, the Colorado Convention Center, and the Denver Museum of Nature and Science, among others) to creation of the new downtown bike-sharing program and of a LEED silver building requirement for city-constructed facilities. City policy interventions affect transportation and land-use design, zoning, and building codes (the FasTracks transit development, Blueprint Denver land-use and transportation guidance, and solar zoning among them). Public venues such as Red Rocks Amphitheatre and the Colorado Convention Center have also been used to showcase zero-waste practices.

Annual report cards track metrics established by the city to monitor progress toward these goals. Institution by the city's Department of Environmental Quality of an environmental-management system citywide has created an infrastructure for standardized reporting so that Greenprint staff can evaluate and adjust programs in a more systematic way. A 2007 report card detailed progress toward the city's emissions goal, and showed the city's efforts paying off, with reductions on track from 2005 to 2007 to meet the long-term goal. Very few other cities (Portland being one notable example) have these types of metrics and reports in place to track progress, despite the widespread adoption of climate goals.

Challenges

Challenges for city-greening programs include building institutional longevity and resources to achieve ambitious goals in the face of changing administrations and budget cuts, as well as continually reassessing the greatest value added for local government's role as state and federal programs and policies evolve, and tracking results for long-term reporting as goals change over time. Maintaining local interest and support for local climate action in the face of a skeptical media and a poor economy can be challenging as well. The recent infusion of federal funding made available to well-positioned cities has actually bolstered greening-program positions in many cases.

Small-city innovations can also serve to incubate policy models for larger cities, and possibly regions. As an example, Aspen's ZGreen carbon-credits program developed a highly credible and well-designed system around 2007 for local residents and businesses (Aspen Skiing Company) to invest in local renewable-energy and energy-efficiency projects while offsetting their own carbon emissions. The program worked so well that it has served as a model for the state of Colorado's carbon-credits program, and its director was hired to run a similar effort for the City of San Francisco. Similarly, when Denver undertook its greenhouse gas inventory, it worked with PhD candidate engineers at the University of Colorado at Denver's Sustainable Urban Infrastructure program under Dr. Anu Ramaswami to develop a new and more comprehensive means of measuring community carbon emissions than the one provided by Local Governments for Sustainability (originally the International Council for Local Environmental Initiatives, or ICLEI) and used by many cities. The results were so successful that the team was invited to participate in revamping ICLEI's standards and has helped to influence urban carbon metrics nationally.

The Role of Cities in a National Context

There are a few cities with sustainability programs where talk about climate change is still verboten (Houston comes to mind), but in general, city-greening programs have also attempted to understand and address local greenhouse gas emissions in some fashion. In past years, this was a direct response to public sentiment that something needed to be done about climate change (in the absence of federal action). The leading role that cities played until just recently in advocating for climate protection, however, has been overshadowed by state and emerging federal policy changes, raising the question of where the value added is in city programs and policy interventions. If price signals for carbon use can have the desired big-picture impacts, why worry about individual carbon footprints? With the new federal focus on getting federal greenhouse gas regulation passed, the pressure on cities to solve the problem is reduced, and the role of city policies in achieving emissions reductions may seem less important.

And yet today, city efforts are more relevant than ever. No matter how sophisticated federal climate policy becomes, the implementation of much of that policy (new building standards, new transportation policies, increased fuel costs, the sale and restriction of carbon permits) will need to take place at the local and state levels, and federal elected officials will need to work closely with their local counterparts to explain new policies to constituents. And, where federal policy makers are just beginning to try to understand where and how to impact various emissions sectors, cities and states in many cases have been looking at related challenges for years and have many lessons to share (with transportation and economic development pilots and green-building policy implementation). Local innovation will doubtless continue to be central to the long-term success of any national climate policy.

Notes

1. In Arizona: Apache Junction, Bisbee, Buckeye, Bullhead City, Flagstaff, Gilbert, Goodyear, Phoenix, Tucson, and Winslow. In Colorado: Aspen, Basalt, Boulder, Carbondale, Denver, Dillon, Durango, Frisco, Glenwood Springs, Gunnison, Nederland, New Castle, Pagosa Springs, Telluride, the town of Crested Butte, and Westminster. In Idaho: Bellevue, Boise, Hailey, Pocatello, Sandpoint, and Sun Valley. In Montana: Billings, Bozeman, Missoula, and Red Lodge. In New Mexico: Alamogordo, Albuquerque, Capitan, Las Cruces, Ruidoso, Santa Fe, and Taos. In Utah: Moab, Park City, and Salt Lake City. In Wyoming: Jackson.
2. See the full plan at www.greenprintdenver.org/about/climate-action-plan-reports.

AFTERWORD

I want to congratulate Beth Conover on her vision for this timely publication, a richly diverse compilation of essays that might seem audacious in its breadth until we, the readers, recognize that climate change is a topic that knows no bounds. Her varied collection about the social, cultural, economic, and political impacts of climate change on our region captures this important moment in history as might a wide-angle lens.

From scientists to evangelicals, from data to dogma, Conover shows us both how and why we should become better stewards of our planet. Over the years, it will become apparent that the timing of this publication squares with the tipping point in almost every one of these varied topics.

Here in the West, the momentum for change has become unstoppable. When I ran for governor in 2006, I made a promise to the people of Colorado to lead our state toward self-sufficiency by promoting alternative energy, encouraging cleaner ways of extracting and using fossil fuels, and rewarding efficiency and conservation. Because this effort was as much about creating a sustainable energy future as it was about creating sustainable jobs, we coined the phrase "the New Energy Economy."

I am happy to report that we are well on our way. We are better utilizing our intellectual capital at our universities and national labs through collaboration. That means we can get the results of important research and development right out in the marketplace. We attracted manufacturing and other economic opportunities by bolstering incentives for the use of renewable energy. We helped make rural Colorado a driver in the expansion of sun, wind, and bio-fuel markets. Homeowners and business owners are actively seeking ways to reduce their energy consumption and their costs. There is demand for more efficient buildings and transportation choices. Our antiquated electricity grid is going to be retooled. The term *carbon footprint* is part of our lexicon.

Soon, the New Energy Economy will no longer be new. It will quickly become our sustainable way of life.

People grapple with complex challenges in myriad ways, and climate change is nothing if not complex. In his seminal book *A Sand County*

Almanac, Aldo Leopold described an "ethic dealing with man's relation to land and to the animals and plants which grow upon it."

> All ethics so far evolved rest upon a single premise that the individual is a member of a community of interdependent parts. His instincts prompt him to compete for his place in that community, but his ethics prompt him also to cooperate...
>
> The land ethic simply enlarges the boundaries of the community to include soils, waters, plants, and animals, or collectively: the land.

I think Leopold might agree it would be appropriate to define a carbon ethic that enlarges the boundaries of the community of cooperation to include climate. An ethic that allows for continued competition but demands cooperation. Where humankind recognizes that only together can we stem the dire consequences of extreme climate change.

In the summer of 2008, I embarked on a tour of the Arctic Circle with world-renowned scientists and climate leaders. It was a once-in-a-lifetime opportunity, and it drove home the realization that we as a nation have the capability, the responsibility, and the opportunity to be global leaders on this issue.

We also owe it to future generations of Coloradans—generations of Coloradans we will never meet—to protect our natural resources, our water supplies, and our crisp mountain air. We have no greater ethical duty.

Westerners are renowned for their resilience, and resilience requires adaptability. Indeed, we must prepare to adapt to some unavoidable changes. But as we face this challenge, it doesn't have to be looked at through a lens of doom, gloom, or desperation. Instead, we can view it with a hopeful sense of possibility and opportunity.

Beth Conover asked an array of leading thinkers to share their observations with us. By doing so, she allows us to see the path forward just that much more clearly.

Thank you, Beth.

—Governor Bill Ritter Jr.
Denver, Colorado

ACKNOWLEDGMENTS

When I first broached the idea for this book with Fulcrum Publishing, I thought it would be interesting and ultimately rewarding. I had no idea (truly) how much fun it would be. Developing this collection has allowed me to contact and work with many of my favorite writers and thinkers in the region and to engage in a creative process with them. I am grateful to all of my contributing authors and artists for sharing their work—most of them original or revised for this volume. Many of them didn't know me, and most took this project on faith and with far more publishing experience than I.

I am grateful to Sam Scinta, who was enthusiastic from the beginning and provided wind in my sails more than once when I doubted its progress.

A few friends (many of them contributors themselves) helped me extend my reach by identifying and/or contacting a wonderful and diverse group of writers. They include Diane Carman, Ted Conover, John Daley, and Heidi MacIntosh, Kirk Johnson, Jeff Lee of the Rocky Mountain Land Library, Laura Snapp, Randy Udall, and Florence Williams. Ted and Florence also served as wonderful informal editors.

Katie Wensuc provided a consistently intelligent, light, and sure touch as a copy editor. Jim Cowles volunteered early editing help with many essays in the book. Anne Button, Brenda Abdilla, Carrie Merscham, and Margo Conover offered valuable feedback on my own writing. Susan Innis, Janet Day, and Pat O'Driscoll shared great ideas and contacts. Michele Weingarden, Sarah Van Pelt, and Scott Morrissey were critical for making sure my knowledge was up-to-date and accurate. My business partner, Matthew H. Brown, both contributed an essay and tolerated my absence from our business start-up while this book was being compiled. GG Johnston offered excellent ideas on sharing the book with a broader audience. Anna Jones; Katie Conover, Jerry Conover, and Jacquelyn Wonder; Pam Conover; and our wonderful neighbors on West Hayward Place all provided critical emotional and moral support. Debra Johnson magically offered a wonderful work space and good company at a critical juncture, and *High Country News* agreed to release a number of previously published pieces for this volume (in revised form). Betsy Kimak also deserves great thanks for her assistance with the website www.howthewestwaswarmed.com.

Denver mayor John Hickenlooper deserves great thanks for helping me to launch a career as a "climate professional" (an odd title but one I wear with pride) and to grow Greenprint Denver, without which I would never have endeavored to take on this topic.

Last, I thank my husband, Ken Snyder, who read and commented on many drafts, offered excellent counsel and support, and never let me take myself too seriously.

CONTRIBUTING ARTISTS

Mike Keefe is the nationally syndicated editorial cartoonist for *The Denver Post*. His drawings have been published in more than 800 newspapers and magazines, including *The New York Times*, *USA Today*, *The Washington Post*, *Time*, and *Newsweek*, and have earned numerous journalism awards.

Kevin Moloney is a freelance photojournalist currently based in Denver, Colorado. He is a regular contributor to *The New York Times*, covering the Rocky Mountain region. His work has also appeared in *US News & World Report*, *Fortune*, *Life*, *Time*, *Stern*, the *Chicago Tribune*, *The Independent*, *USA Today*, *Elle*, *Marie Claire*, *Business Week*, *The Christian Science Monitor*, and National Geographic publications. He was one of two journalists selected as inaugural recipients of the Ford Fellowships in Environmental Journalism.

Anne Sherwood is a freelance photojournalist based in Bozeman, Montana. Her work appears frequently in *The New York Times*. Although she's never been known to turn down a plane ticket, Sherwood finds as much inspiration in her own backyard as she does on the other side of the planet.

Ed Stein is the former editorial cartoonist for the *Rocky Mountain News*. His work is syndicated internationally by United Media.

ABOUT THE EDITOR

© Susan Goddard

Editor Beth Conover has worked for twenty-five years at the intersection of environmental protection and economic development. As policy advisor to Denver mayor John Hickenlooper, she was the architect of Greenprint Denver, one of the nation's earliest and largest urban sustainability programs, and helped lay the groundwork for the greening of the Democratic National Convention in 2008. She is a graduate of Brown University and holds a joint MBA/masters of environmental studies from Yale University. She is the recipient of the Mitchell Young Scholar Award for Sustainable Development and the Wirth Chair Sustainability Award of the University of Colorado at Denver. Conover is a founding partner in the consulting firm ConoverBrown, LLC. She is a native of Denver, Colorado, where she lives with her husband, Ken Snyder, and their two sons.